CW00374970

Where to Watch Birds in

# THE WEST MIDLANDS

OTHER *WHERE TO WATCH BIRDS* BOOKS PUBLISHED BY
CHRISTOPHER HELM

*Where to Watch Birds in Devon and Cornwall*
David Norman and Vic Tucker

*Where to Watch Birds in East Anglia*
Peter and Margaret Clarke

*Where to Watch Birds in Bedfordshire, Berkshire, Buckinghamshire, Hertfordshire and Oxfordshire*
Brian Clews, Andrew Heryet and Paul Trodd

*Where to Watch Birds in Kent, Surrey and Sussex*
Don Taylor, Jeffery Wheatley and Tony Prater

*Where to Watch Birds in Wales*
David Saunders

# Where to Watch Birds in
# THE WEST MIDLANDS

*Including Hereford-Worcester, Shropshire, Staffordshire,
Warwickshire and the former West Midlands County*

Graham Harrison and Jack Sankey

CHRISTOPHER HELM
London

© 1987 Graham Harrison and Jack Sankey
Line illustrations by Brett Westwood. Maps by Janet Harrison and the authors
Christopher Helm (Publishers) Ltd, Imperial House,
21–25 North Street, Bromley BR1 1SD

British Library Cataloguing in Publication Data

Harrison, Graham R.
    Where to watch birds in the West Midlands:
    including Hereford-Worcester,
    Shropshire, Staffordshire, Warwickshire
    and the former West Midlands County. —
    (Where to watch).
    1. Bird watching — England — Midlands —
    Guide-books   2. Midlands (England) — Description
    and travel — Guide-books
    I. Title   II. Sankey, Jack
    598'.07'234424      QL690.G7

    ISBN 0-7470-1402-7

Typeset by Opus, Oxford
Printed and bound in Great Britain by
Biddles Ltd, Guildford and King's Lynn

# CONTENTS

# ACKNOWLEDGEMENTS

We have received considerable help from several people, but particularly wish to express our gratitude to Brett Westwood for his excellent illustrations and to Janet Harrison for drawing so many of the maps.

We are grateful to Leo Smith who helped with the Shropshire sites, and to the following for their expert advice and constructive comments: Michael Boote and his 'team' for Doxey, Alan Dean for Blithfield, Graham Evans for Chasewater, Frank Gribble for many Staffordshire sites, Dave Smallshire for Belvide, Steve Welch for Sandwell Valley and Steve Whitehouse for Worcestershire. We are also grateful to the staff of the County Naturalists' Trusts for their help and assistance. Finally we should like to thank the staff at Christopher Helm Limited, without whose help and guidance this book would not have been possible.

Graham would also like to thank his wife, Janet, and daughters Susan and Jennifer for their help in typing drafts and for their encouragement and tolerance when the task took over their lives, and Jack would like to thank his parents for their tolerance and understanding of his neglect whilst he was working on this book.

Graham R. Harrison

Jack Sankey

# INTRODUCTION

This book is a guide to birdwatching sites in Hereford-Worcester, Shropshire, Staffordshire, Warwickshire and the former West Midlands County. Few of these are likely to figure prominently in any national list of prime haunts. Yet to dismiss such a varied and interesting region out-of-hand is as unwise as it is unjust.

The chief asset of this area, which we shall call the West Midlands, is its position astride the great national divide between uplands and lowlands. It is a region of transition, both from the bleak, sombre moors of the Peak District to the soft, mellow hills of the Cotswolds; and from the babbling streams of the Welsh border country to the lush, fertile valleys of the Avon, Severn and Trent.

Within it is a rich mosaic of habitats that supports northerly species like Twite and Black Grouse, southerly ones like Nightingale and Hobby, westerly ones like Pied Flycatcher and Buzzard, and easterly ones like Red-legged Partridge and Corn Bunting. The hub of the region is Birmingham, where Black Redstarts add to the diversity of birdlife and with the Marsh Warbler in Worcestershire the region has a national rarity.

The main drawback is undeniably the lack of coastline, which limits the number of species likely to be seen and the propensity for rarities to occur. Yet vagrant seabirds do occur with some regularity, often affording better views than in their more natural surroundings.

The purpose of this book is, then, to help birdwatchers to enjoy the wealth of interest to be found in the West Midlands by showing them not only where to look for birds, but when to look. It is intended both for newcomers and experts; for residents of the West Midlands and for visitors to the region.

In compiling the book we have concentrated on: (1) Sites which regularly hold good numbers or a good variety of species. These can generally be relied upon to be productive and are worth visiting even when time is limited. (2) The best examples of each habitat. These are worth exploring for specific communities or species, but they may lack all round interest and require more time than (1) to make a visit really worthwhile.

We have avoided private sites and concentrated on those to which the public can gain access. We have also concentrated on stable sites, omitting some that are currently interesting but likely to change. Overall, the sites we have chosen are amongst the best in the region, but our list is by no means exhaustive and many a good days' birdwatching can be enjoyed elsewhere. Indeed, we would be the first to encourage people to explore, since there is no greater reward in birdwatching than making a discovery of your own. For those with limited time or knowledge, however, our selection should help to make the most out of visits.

We trust everyone will understand the need to withhold certain information. Some sites are too private and inaccessible to mention, while others are susceptible to damage or disturbance by people. Above all there is the welfare of the birds themselves to think of. We have meticulously avoided disclosing sensitive or confidential breeding or roosting sites. At the same time, we have tried to suggest where every

1

typical species might be seen. On occasions this has meant disclosing spots where sensitive species can be watched from public rights-of-way. We have done so in good faith: firstly in the hope that it will discourage indiscriminate and potentially even more damaging random searching, and secondly in the hope that all our readers will act responsibly and not create disturbance by seeking closer views. We ask you please to repay this faith by visiting, watching and leaving the site undisturbed for others subsequently to enjoy.

# HOW TO USE THIS BOOK

There are innumerable field guides to help identify birds. Yet strangely there are few on where, when and how to find birds and those guides that do exist are often very superficial. To read beforehand about all the exciting birds to be seen somewhere, only to find the place deserted when you arrive is very frustrating. Scanning across an empty reservoir or discovering you have several square miles in which to search for a particular species can be very disheartening. Immediately you ask how many birds are there, where should I look first and when should I look?

To answer such questions we introduce the reader to a cross-section of the region's better birdwatching sites. Each site is then dealt with systematically, with sections on habitat, species, timing and access, followed by a calendar.

## Habitat

This section aims to provide the reader with a mental picture of each site by describing both natural and man-made features and the range of vegetation to be found. We make no excuse for having described some sites in detail, as a thorough knowledge and understanding of habitats makes it that much easier to find the birds.

## Species

This section begins with a statement of the main ornithological interest of each site. Following this is an account of the species to be found. Space has not permitted a comprehensive account, but in any case it is not our intention to provide a complete list of every bird ever seen at a site. Rather we have set out to give the reader an idea of the more interesting birds to be found. Thus the commoner species have normally been omitted and rare or unusual species are mentioned only as examples of what might occur. Visitors should not expect to see these specific species.

Most species' accounts have been arranged to give an idea of how the birdlife changes through the year. They also indicate why the birds are using the site. For example, they might be breeding, feeding or roosting; or they might be passing through or arriving during periods of flood or extreme cold. Where possible an indication is given of how often the birds occur and in what numbers, so that visitors know whether to expect a single bird, small party or large flock. Numbers and frequency have generally been based on the period 1981–5, which for convenience has been referred to as 'recent'.

## Timing

Knowing when to visit is usually more important for coastal than inland sites, as there are no tides or onshore winds to worry about at the latter. In a densely populated area like the West Midlands, though, disturbance can be an important consideration. At times even the sheer number of birdwatchers can cause problems, with hides full to overflowing. Most disturbance is caused by recreational activities, particularly water-based sports. While we have attempted to assess the effects of this as accurately as possible, readers should bear in mind that activities like these are always liable to change at short notice. Invariably they intensify rather than reduce. Weather is the other important factor. Wind direction is not usually that significant, but periods of flooding and severe frost are.

## Access

Directions are given to every site from the nearest main road. In some cases there are alternative approaches, but to save space only one has been described. The alternatives can usually be followed by using the relevant site map (at the back of the book) and/or by reference to the Ordnance Survey map; numbers for both are given against each site heading throughout. Parking arrangements are described and details of footpath access are given where this is thought to be relevant. (This section should always be read in conjunction with the site maps.)

Where access to a site is strictly by permit only this is clearly stated and visitors are asked to respect this. Trespassing only serves to get birdwatchers a bad name and can lead to the privilege of access by permit being withdrawn altogether. In the case of certain County Trust sites, access is also available to members of other trusts, but this should be verified with the respective trust office beforehand.

## Calendar

This provides a quick reference to the birds likely to be seen at different times of the year. Resident birds are those that might be seen at any time of year, though they are not necessarily always present. Most also breed, though again this is not necessarily the case. For example, species like Grey Heron regularly feed at reservoirs, but may nest some miles away. There are three periods relating to the main passage or wintering times: spring is defined as March or April until June, which is when most waders pass through and migrants arrive; autumn is defined as July to September, which is when most summer visitors depart and waders again pass through; winter is classed as October to March or April.

If there is no qualifying comment, the bird concerned might be expected at any time during the period for which it is mentioned. The terms 'scarce' and 'rare' refer to frequency of occurrence rather than total numbers. To find when peak numbers are most likely, refer to the species section for each site. Birds, like all living things, are classified or grouped according to their taxonomic differences. Classification is constantly evolving with new knowledge and the sequence of species changes. We have chosen to use the Voous Classification, which is that followed by most contemporary field guides.

We have tried to ensure that your visit will be both enjoyable and rewarding. But birds undertake complex movements in response to a whole range of unpredictable circumstances. Inevitably there will be times when a site fails to live up to its expectations and we would ask for your forebearance in this. Hopefully such occasions will be few and offset by those when a site exceeds its expectations.

# A PORTRAIT OF THE WEST MIDLANDS

To most people the West Midlands means Birmingham or motor cars. Yet there is much more to it than that. Its five constituent counties cover 5,000 square miles (13,000 km$^2$) of extremely varied countryside that ranges in altitude from 40 to 2,306 ft (12 to 703 m). The climate is varied too. The effect of the Gulf Stream means relatively mild winters and warm summers in the Avon and Severn Valleys. Spring is always a week or two earlier here and this climatic advantage probably influences the distribution of birds such as Nightingale and Marsh Warbler. Although the mean January temperature is above freezing everywhere, this disguises the propensity for extremely cold nights and prolonged frosts. The Peak District also experiences heavy snowfalls. East of the Severn the region is surprisingly dry, with less than 30 in (750 mm) of rain each year. This favours arable farming and those birds associated with it. West and north of the Severn, rainfall increases steadily until it exceeds 40 in (1,000 mm) in the Black Mountains and the Peak District.

Of course weather and topography very much influence bird migration. Inland movements are never as spectacular as those on the coast, but long-distance migrants travelling, say, between their Scandinavian breeding grounds and African wintering grounds will readily overfly the British Isles. There is also considerable movement between the east and west coasts, particularly between the Wash and the Severn Estuary. If birds encounter adverse weather conditions whilst making such movements they may well drop into the Midlands to feed and rest. During autumn storms, westerly winds often drive one or two seabirds well inland, while in winter cold weather in the North Sea and the Baltic may well bring movements from the north and east.

To journey across the region from south-east to north-west is to step backwards in time. Beneath your feet the rocks get progressively older, whilst around you man's influence on the landscape declines and nature's begins to dominate.

The newest rocks are the Jurassic beds of the Cotswolds. They are followed by the Trias of the great Midlands Plain and then by the Carboniferous deposits, which include both the coal measures that kindled the Industrial Revolution and the grits and limestones of the Peak District. Further west, Old Red Sandstones underlie most of Herefordshire, and finally the oldest rocks are the Silurian, Ordovician, Cambrian and pre-Cambrian beds that form the backbones of the Shropshire Hills. Within this generalised pattern are localised intrusions of igneous and metamorphic rocks, most notably the Malvern Hills.

The scenery reflects this varied geology. More importantly, the bedrock has strongly influenced man's activities. Thus, in the south and east relatively few semi-natural habitats survive and the best birdwatching sites tend to be man-made ones like reservoirs. By contrast, in the north and west it is the characteristic habitats that are important, rather than particular sites.

Standing atop the great Oolitic scarp of the Cotswolds — at Edge Hill, Broadway Hill or Bredon Hill — you are surrounded by warm, honey-coloured stone walls and picturesque villages. Before you stretches the great, flat Midlands Plain. In the foreground are the heavy

Lias clays of south-east Warwickshire and Worcestershire, the arable and horticultural centre of the region. Here, fields of wheat, barley and vivid oilseed rape stand alongside the old ridge-and-furrow of heavy pastures. So tidily is this area farmed that precious little land is left to nature and the only birdwatching site of any importance is the man-made Draycote Water. True there are one or two limestone quarries, such as that at Ufton Fields, and a few small woods, but for the most part birdwatching here is extremely hard work. Flocks of Lapwings and Golden Plover on unimproved pastures in the winter, breeding Hobbies in the summer, and autumn flocks of passerines following the Cotswold scarp are the main interest. Around Evesham, orchards, greenhouses, polythene cloches and the other paraphernalia of market gardening and horticulture proliferate. Birdlife here is quite impoverished compared with that of 50 years ago, when the gnarled old orchard trees held many hole-nesting birds.

Separating the Jurassic from the Triassic is the first of the region's four great rivers — Shakespeare's historic Avon. Dotted throughout its valley are charming half-timbered villages, grand country houses and parklands whose subtle blend of grasslands, lakes and woods supports very many of the commoner birds. In winter the flood meadows just upstream of Tewkesbury hold wildfowl and waders including quite often a herd of Bewick's Swans. The same area was once the national stronghold of the rare Marsh Warbler, and despite a recent collapse in numbers a few pairs can still be found in summer. Much further upstream the river is flanked by extensive gravel-bearing terraces, which here and there have been quarried for aggregates. Ornithologically, the best of these quarries is Brandon Marsh, which is excellent for wintering wildfowl, passage waders and unusual marshland birds.

The bulk of the vast Midlands Plain lies north of the Avon and is underlain by beds of Keuper marl with a variable covering of glacial drift. This is mixed farming country, with picturesque half-timbered cottages and elegant red-brick houses. Fields are still reasonably small, hedgerows well timbered, and small woods are liberally scattered throughout. Some of these woods are reputedly remnants of the once extensive old hunting forests of Feckenham and Arden, though invariably much altered by man. Across the Keuper beds pedunculate oakwoods predominate and until the Second World War most of these were regularly coppiced. This practice produced ideal conditions for Nightingales and scrub warblers, and even today the area around Himbleton remains the centre of a much depleted Nightingale population. In north Warwickshire the marls are replaced by sandstones and in woods such as Bentley Park, the more acidic soils support both species of oak and a few birds such as Redstart and Wood Warbler that are typical of true sessile oakwoods. One site of special importance is Upton Warren, where brine seepage into shallow subsidence pools creates one of the best wader habitats in the region.

Within the Midlands Plain are three outcrops of Carboniferous coal measures that have had a profound effect on the development of the region. It is these that cradled the great industrial centres of Birmingham and the Black Country, Coventry, and the Potteries. Today these areas house nearly three million people. Needless to say even semi-natural habitats are few in this maze of concrete, bricks and mortar, but industrialisation has brought new habitats for our wildlife to exploit. In particular, large industrial installations have been taken over by that

6

urban speciality the Black Redstart. Further out, the leafy suburbs now provide safer homes for many common passerines than do their more customary rural haunts. Birmingham once stood at the hub of a thriving canal network. To replenish this network several feeder reservoirs were constructed, of which Bittell and Chasewater are the best for birds. Add to these, water-supply reservoirs like Bartley and the new river purification lake at Sandwell Valley and it is clear that the core of the region is now much more attractive to wildfowl, waders and gulls than ever before.

Coventry is a much newer city, substantially rebuilt since the Second World War blitz. Although well endowed with open spaces, these have yet to prove really attractive to birds and the best birdwatching sites lie just beyond the city. Whilst Birmingham and Coventry grew progressively outwards from their centres, engulfing many small villages as they went, the Black Country towns like Dudley, Stourbridge, Walsall, West Bromwich and Wolverhampton developed simultaneously, leaving isolated pockets of countryside between them. Today these pockets bring precious countryside within easy reach of a vast population. They also support an unexpected variety of birdlife, from Wood Warblers in the tiny Saltwells Wood to wildfowl and waders in the Sandwell Valley. Some 40 miles (60 km) to the north, the 'five towns' of the Potteries, which Arnold Bennett so vividly described, grew in much the same fashion as did those of the Black Country. Today though, Stoke-on-Trent is very different. Gone are the bottle kilns, collieries, foundries and slag heaps. Now the emphasis is on reclamation and many of today's better birdwatching sites were derelict wastes not so long ago.

Modern urban society has extended its influence well beyond the built-up areas of our towns and cities. Nowhere is this more evident than to the east of Birmingham, in the Tame Valley. Here flood meadows, gravel pits, colliery flash pools, a reservoir, power station, river purification scheme and water park form a unique chain of wetlands that are of outstanding importance to birds. Amongst the surprising array of species are Shelduck, Oystercatcher, Ringed Plover and Common Tern — all nesting right in the very heart of the country.

From Birmingham and Kidderminster northwards the Keuper marls of the river basins are enclosed by ridges of higher ground with tracts of heathland on the poor, acidic soils derived from the underlying Bunter pebbles and sandstones. Sutton Park is a fine example, which despite being totally surrounded by houses, supports a good population of heath and woodland birds. Even better is Cannock Chase. Here there is a superb mosaic of dry and wet heath intermingled with both broad-leaved and coniferous woodland. The birdlife is extremely good, with most heath and woodland species present, including Crossbill, a good population of Nightjars, and usually a Great Grey Shrike in winter. Similar heaths occur to the west of the conurbation, at Kinver and Enville, and to the south of the Potteries, between the Churnet Valley and Hanchurch Hills. Most have been afforested, but their birdlife, though interesting, is not as rich. Bunter beds also occur in south Staffordshire and east and north Shropshire, but here they are often covered by a coarse glacial drift which supports fields of barley, wheat, potatoes and sugar beet. In winter such areas often hold flocks of Golden Plover.

North of Cannock Chase, the flat plain of Keuper marl continues to the Potteries and beyond into Cheshire. In the distance is the outline of the Peak District while in the foreground the region's second great river, the Trent, glides peacefully along. This is often said to be Britain's great divide — separating northerners from southerners and uplands from lowlands. Its influence on the region's birdlife is similar, with for example, good numbers of Goosander here, whereas they are much scarcer further south.

The plain, once again, is dairy farming country, with lush pastures enclosed by thick hedges. Though not the richest part of the region for birds, it does possess several very important wetlands, both natural and man-made. In south Staffordshire the canal feeder reservoir at Belvide, the water supply reservoir at Blithfield and the subsidence pools at Doxey Marshes are all excellent places for wildfowl, waders and gulls. Indeed, Belvide probably holds the highest density of wildfowl in the region, while the larger Blithfield Reservoir is one of the foremost sites in the country for wintering wildfowl. It is also easily the best site in the region for passage waders. In Shropshire interest centres on more natural sites, particularly the glacial meres around Ellesmere which hold good numbers of wintering wildfowl, and mosses such as Wem and Whixall.

Across northern Staffordshire glacial meltwaters have cut a magnificent gorge through which now flows the River Churnet and its tributaries. The valley is clothed with mixed woodlands, including some superb sessile oakwoods such as those at Coombes Valley. Here characteristic upland species like Redstart, Wood Warbler and Pied Flycatcher can be found, while the fast-flowing streams are the haunt of Dipper and Grey Wagtail. At the head of the Churnet Valley, the twin reservoirs of Rudyard and Tittesworth hold wintering wildfowl and passage waders. This is also one of the few areas in the region where Whooper Swans regularly winter.

Beyond the Churnet the land rises still higher to the plateaux and craggy summits of the Peak District, which in places stand above 1,500 ft (500 m). This is an area of two quite distinct parts. To the east, white drystone walls and sturdy, bright farmsteads mark the Carboniferous limestone. Here the plateau is dissected by the majestic dales, with their crystal streams, ashwoods and sheer rock faces. This too is an area for Redstart and Pied Flycatcher in the woods, and for Dipper and Grey Wagtail along the rivers. To the west, on the Millstone Grit, are extensive tracts of windswept grass and heather moor, interspersed with upland pastures, rocky crags and deep, wooded cloughs. This is an outstanding area for breeding birds. Most birdwatchers come to see Red and Black Grouse, Golden Plover, Wheatear, Ring Ouzel, Twite and passing raptors, but there are also important populations of Lapwing, Curlew and Snipe.

Returning to Birmingham, a hilly ridge extends north-westwards from the Lickey to the Clent Hills. This provides, within easy reach of the conurbation, a range of habitats from open hillsides to mature mixed woodland. The Lickeys hold a good variety of woodland birds and are well known for their winter Bramblings, while the more open hillsides of Clent are sometimes visited by passage Ring Ouzels.

Further west lies the region's third great river, the Severn. Just as the Trent divides the country, so the Severn divides west from east and upland from lowland. Through Shropshire it meanders across a broad flood plain, where the meadows upstream of the spectacular Ironbridge Gorge hold wintering wildfowl in times of flood. Often these include a few Bewick's or Whooper Swans. On the plateau above the gorge is the region's fourth outcrop of Carboniferous coal measures. Although only a small coalfield, Ironbridge became famous throughout the world as the birthplace of the Industrial Revolution. Now it is experiencing another renaissance as Telford New Town.

Once through the Ironbridge Gorge the Severn follows a more direct course down to the Bristol Channel. On its western bank just below Bridgnorth lies Chelmarsh Reservoir, which is one of the foremost sites in Shropshire for wildfowl. Just a few miles downstream, on the same bank, the Wyre Forest is one of the finest and most extensive forests in Britain. Here can be found an excellent range of woodland species including Redstart, Wood Warbler, Pied Flycatcher, Crossbill and Hawfinch. On the opposite bank, to the south of Kidderminster, lowland heaths are a feature of the landscape. These are of some ornithological interest, with perhaps breeding Stonechat; but, sadly, species like Nightjar, Woodlark and Red-backed Shrike belong to years gone by.

West of the Severn the landscape becomes markedly more hilly. In the south, some of the oldest pre-Cambrian rocks in Britain rise abruptly from the plain beneath to form the distinctive outline of the Malverns. The lower slopes hold Wood Warblers and are often visited by Ring Ouzels in autumn, while the bare summit grasslands are the haunt of Wheatears in summer and sometimes Snow Bunting in autumn. The hills also attract passing raptors. The tail of the Malverns extends northwards as a broken ridge of Silurian rocks for some 12 miles (20 km), sometimes with open summits but usually under woodland. Here Dippers and Grey Wagtails inhabit the fast-flowing streams and Buzzards soar above the woods.

From the summits of the Malverns the views are spectacular. To the east is the Midlands Plain, while beyond and to the south-east the prominent scarp of the Cotswolds recedes into the distance. To the west lie the rolling hills of Herefordshire, which must surely be one of England's loveliest and least spoilt counties.

Herefordshire is a bowl of red sandstone, traversed by broad, fertile river valleys and ringed by hills. To the east are the Malverns, to the south the Doward and the Forest of Dean, to the west the Black Mountains and the Welsh Hills, and to the north the Shropshire Hills. Within the bowl are lower ridges of sandstone and a dome of Silurian limestone around Woolhope that is the centre of some superb, wooded countryside. Many woodland birds are to be found in this district, including Crossbill and Nightingale, which here is just about at its most westerly limit. This is truly a farming county, noted for its rich, red soils and famous red-and-white cattle. The landscape is one of corn and beet fields, lush pastures, hop-yards and cider orchards. Narrow, twisting lanes link charming, half-timbered villages to the small market towns. Woodland abounds, particularly on the steep hillsides. On the sandstones oak is the native tree, but beech flourishes on the limestones and both alder and willow are common along the watercourses. There is plenty of holly, especially in roadside hedges, and clumps of mistletoe are commonplace. Many of the woods have now been replanted with conifers, particularly in the Mortimer Forest along the Shropshire border. This has helped to diversify the avifauna by providing a home for Siskin, Redpoll and Crossbill. None the less there is still sufficient broad-leaved woodland for birds like Redstart and Wood Warbler to be reasonably common, for Hawfinches to occur here and there and for Pied Flycatcher to be the typical bird of Herefordshire.

Through the middle of the bowl flows the fourth great river of the region, the Wye. The Wye Valley must be one of the most beautiful in England, particularly where it flows through the magnificent, wooded gorge at Symonds Yat. Here the spectacular beauty of the scenery is matched only by the breath-taking agility of the resident Peregrines. Freedom from pollution and the absence of standing water make the Wye important to birds. Indeed it is the principal haunt of wildfowl in Herefordshire. Small numbers of wintering ducks and passage waders can be found almost anywhere along its course, but there are particular concentrations around the Lugg confluence and downstream of Hay-on-Wye. Winter flooding in these places often brings increased numbers of dabbling duck, especially Wigeon and Teal, and the possibility of a few wild swans.

In the extreme west the Old Red Sandstones rise to the plateau of the Brecon Beacons, which at over 2,300 ft (700 m) is the highest point of the region. The rocky crags and scree slopes of these lofty summits are the summer haunt of Wheatear and Ring Ouzel, while on the plateau the windswept heather moor holds Red Grouse and perhaps Merlin. Down below, the secluded Olchon Valley holds the typical species of upland streams and woods. North of the Wye the Silurian hills of Radnorshire close in around Kington. Nestling in the lee of these hills is the tiny Eywood Pool, which holds a few winter wildfowl.

The Silurian hills then continue across to Ludlow and into south-west Shropshire, where they are heavily dissected by numerous streams and rivers. Here rough hill grazing and lush valley pastures combine to produce an unspoilt landscape in which there are more sheep than people. This is the countryside of wooded hillsides and sheepwalks that are the haunt of Buzzard and Raven. Afforestation in the Clun Forest has introduced alien conifers and their associated birds into the landscape. Further east the hills develop into parallel ridges along a south-west/ north-east axis. The longest of these ridges is Wenlock Edge, where the mixed woodland holds a good variety of birds. Of even greater interest though, are the older rock outcrops to the north and south. Southwards are the twin peaks of the Clee Hills, where igneous intrusions give rise to acid heath and heather moor that holds Red Grouse and various breeding and passage birds. Northwards are the igneous rocks of the Wrekin with its wooded slopes, and the very ancient pre-Cambrian rocks of the Long Mynd and Stiperstones. The latter two support superb examples of heather moor and rocky outcrops, where Red Grouse and Buzzard are resident and summer brings Wheatear, Ring Ouzel and Merlin.

In this brief outline we have attempted to give a flavour of the region and its birdlife. The following chapters develop this by describing the better sites in much more detail. Even now, though, it should be apparent that the West Midlands has much to offer the birdwatcher. Where else are upland and lowland, northerly and southerly, or easterly and westerly species brought into such close proximity? Add to this a good range of passage birds and the chance of a rarity here and there and the prospects for a successful day's birdwatching are good.

The southernmost part of the Peak District National Park protrudes into Staffordshire, where it embraces two quite distinct habitats. To the east, where the Carboniferous limestone outcrops, are the famous dales, of which majestic Dovedale is the best known. To the west, on the Millstone Grit, are the bleak and sombre North Staffordshire Moors.

## THE DALES

### Habitat

The eastern side of the National Park in Staffordshire is part of a great limestone dome that forms a plateau of short, but nutritious, swards that are grazed by cattle and sheep. Fields are enclosed by drystone walls of white limestone and trees are few, although some farmsteads do nestle behind a sheltering belt of sycamores.

The glory of this area is the dales themselves – spectacular ravines with crystal-clear rivers, a mantle of ash woodland or thorn scrub, and towering limestone cliffs, many of them in the ownership of the National Trust (NT). Dovedale, the eastern flank of which is in Derbyshire, is the narrowest and hence most spectacular. It is also the most popular and throughout the summer is thronged with visitors. Interesting birds can be seen here, but most birdwatchers prefer the Hamps or Manifold Valleys.

The Manifold below Wetton Mill is typical dales country and can be easily explored from the footpath along the old railway track. Just below the mill is a swallet hole down which the river disappears in summer, leaving a dry bed for several miles. In winter, though, it flows throughout its course.

In spring and summer, the small riverside meadows are dotted with the delicate flowers so typical of limestone country. Rising abruptly from the valley floor are steep slopes of grassland or ash woodland. On the upper slopes these give way to gorse or thorn scrub, whilst the summits are marked by sheer cliffs or crags of bare limestone, with here and there natural caves. In places, vast quarries disfigure the landscape.

### Species

Breeding woodland and waterside birds in superb scenery are the main interest.

In early spring the plateau tops come alive with tumbling Lapwings, the bubbling calls of Curlews and the outpourings of Skylark. Less obtrusively Meadow Pipits perch on drystone walls or overhead wires, joined later by a few Wheatears and Whinchats.

In the dales the rivers are the home of both Dippers and Kingfishers. The latter are found on the deeper, slower reaches, particularly where there are overhanging banks for nesting. During a dry season, birds can often be seen catching fish that are trapped in the small pools left behind when the river disappears underground. For Dippers look where there are boulder-strewn shallows or shingle banks for feeding. During the breeding season they may also be seen flying to and from their nests beneath a bridge or the overhang of a bank. Pied and Grey Wagtails also frequent the riverside, especially in the vicinity of mills and waterfalls.

Amongst the ashwoods Chaffinch and Willow Warbler are the commonest species, but Blackcap and Garden Warbler, Spotted Flycatcher and the occasional Green Woodpecker, Marsh Tit or Nuthatch can also be seen along with the commoner woodland birds. Wood Warblers can be found in one or two places and Pied Flycatchers are occasionally seen. Where the closed woodland opens out into scattered trees and scrub listen for Tree Pipits or watch for the fiery red flash of a Redstart's tail. Goldfinch, Linnet, Whitethroat and an occasional Whinchat also occur amongst scrub on the upper slopes, with Linnet and Whitethroat especially associated with the patches of gorse.

Overhead, Rooks are common, Woodpigeons and Stock Doves fly to and fro across the valley and noisy flocks of Jackdaws spiral around the exposed crags, caves and quarries. Swifts are regularly overhead in summer and sometimes nest behind the fissures in the limestone. Likewise some House Martin colonies may be found on the exposed crags. Kestrels, frequently seen hovering or circling into a thermal, are the commonest raptor.

Most summer visitors leave during August and there is little bird activity until late September or October. Then flocks of Fieldfare and Redwing relish the berries on the thorn scrub. They stay so long as the food lasts, then move on leaving the resident birds alone once more.

## Timing

Early morning, when song birds are most vocal, is the best time to visit the woods. The dales get very crowded on warm, sunny days, particularly at weekends and Bank Holidays, forcing many species of birds into remote inaccessible spots. Even Dippers and riparian birds like Pied and Grey Wagtail can be elusive when there are crowds of people about.

## Access

Dovedale is best approached from the south, via the A515 Ashbourne to Buxton road. Leave Ashbourne and 1 mile (1.6 km) out turn left onto a minor road and follow the signs to Thorpe. Bear right in the village, continue down the hill and across the river bridge, then turn immediately right past the Isaac Walton hotel. Park in the car park on the right and follow the main footpath along the river to Milldale, a distance of 3 miles (5 km).

From the north, coming from Buxton, turn right off the A515 about 4 miles (6.5 km) south of the junction with the A5012 and take the narrow lane signposted Alstonefield. At the bottom of the steep hill, cross the river and take the left turn to Milldale. Park in the village (space is limited, so this can be difficult at busy times), cross the river and follow the footpath southwards.

To visit the Manifold continue from Milldale (or through Alstonefield) to Hopedale, then take a right turn signposted Wetton. Park on the roadside by the Y junction at the western end of the village, follow the farm track for a short distance, then take the footpath on the right to reach Thor's Cave and spectacular views along the valley. Check the thorn scrub hereabouts for chats and warblers. Return to the road and drive on down the steep hill to the river. Park near Wetton Mill and follow the footpath southwards along the disused railway for woodland, scrub and waterside species.

Wetton Mill can also be reached from the B5053 through Butterton. The old railway may be followed as far as Weags Bridge, 2 miles (3 km), then onwards up the Hamps Valley to the A523 at Waterhouses, a total distance of 6 miles (9.5 km). Alternatively, explore from the riverside car park at Weags Bridge (to reach this carry straight on from Alstonefield towards Grindon instead of turning right into Wetton).

## Calendar

*Resident:* Kestrel, Stock Dove, Kingfisher, Green Woodpecker, Pied Wagtail, Dipper, Marsh Tit, Nuthatch, corvids and common woodland passerines.

*March–June:* Lapwing, Curlew, Skylark, Meadow Pipit and Linnet (a few of the last three may be resident). Summer visitors from mid-April: Swift (not before May), House Martin, Tree Pipit, Grey Wagtail, Redstart, Whinchat, Wheatear, Whitethroat, common woodland warblers including Wood Warbler, Spotted Flycatcher (not before May) and Pied Flycatcher.

*July–September:* Mostly quiet in July, then summer visitors begin to leave in August.

*October–February:* Winter thrushes return in October.

## THE NORTH STAFFORDSHIRE MOORS
### Habitat

The main areas of moorland lie between Leek and Buxton, reaching a height of 1,684 ft (513 m) on Oliver Hill. Although this is not a great altitude, the climate is most inhospitable; rainfall is heavy and snow frequently lies in winter. By moorland standards the area is well populated, with numerous scattered farmsteads and villages of which Flash, at 1,500 ft (460 m), is reputedly the highest in England. For moorland, it is also surprisingly accessible, being crossed by the A53 and numerous minor roads. Beware in winter though, as roads are frequently blocked by drifting snow.

The land west of the A53 is generally higher, with rocky outcrops, steep-sided valleys and fast-flowing streams. The Roaches and Ramshaw Rocks stand up as exposed sandstone crags, smoothed and contoured by centuries of rain, wind and frost. Below is a bleak, rolling landscape of heather, cotton grass, bracken, rushy pastures and peat bogs, relieved only in the valleys where drystone walls enclose the upland sheep pastures. Here and there is an occasional patch of scrub woodland, usually birch, willow or pine, or a small conifer plantation. Sessile oakwoods occur in the valleys.

### Species

There is an outstanding range of moorland species, especially in summer.

Seasonal contrasts are most marked. Severe winters force all but the hardiest species to forsake the moors, with even Robins sometimes deserting the higher woods. Only the grouse and a few Meadow Pipits, corvids and Reed Buntings can be regarded as truly resident.

Much of the remaining heather is managed as grouse moor and there is a healthy population of Red Grouse. They are most easily seen in autumn, but their presence at any time is often revealed by their harsh 'go-back, go-back' calls. Small coveys might burst unexpectedly from cover in almost any patch of heather, but beware of game birds 'whirring' low across the moor and disappearing into heather again — there are Partridge on the moors too!

*Black Grouse still survive on the moors*

Unfortunately the same cannot be said of the Black Grouse as of the Red. Its population is in decline — so much so that its future in this part of the Peak District is now in jeopardy. (For this reason we are unable to disclose its precise whereabouts.) There are perhaps only a dozen males left and they congregate at their lek to display to the females. Males may return to their leks early in the year, often in February, but activity is seldom intense until the spring. It then subsides quickly after May. Most activity takes place early or late in the day, the most likely times to see these birds as they fly to or from their lek. Their flight is normally much higher than that of the Red Grouse, and the males are readily identified by their tail shape. Sometimes lekking occurs in autumn too, but only in the early morning. Just occasionally a Black Grouse might be encountered in the middle of the day, perched on a drystone wall or even in a tree. Bilberry is a favourite food, and the higher edges of scrub woodland, where bilberry and heather form part of the ground flora, are always worth a look. In winter, the birds may be seen feeding in birch scrub and sallow thickets.

By March, Meadow Pipits begin to reappear and soon they are the commonest, most widespread breeding species. Later in the spring they will become involuntary hosts to a good many Cuckoos. Curlew also begin to return in March, and by April they too are widely distributed and their evocative, bubbling calls are one of the most characteristic sounds of spring and summer. Large flocks of Redwing and Fieldfare may also be seen at this time as they begin their journeys northwards.

Winter lingers on the moors and snow is often still lying when the first Ring Ouzels and Wheatears return in late March or early April. Both species quickly pair and take up residence. Ring Ouzels, of which there are over 50 pairs, seemingly prefer the zone where upland pastures give way to open moorland, particularly where it is strewn with boulders or scree and is close to running water. Drystone walls, derelict barns and scree slopes are favoured by Wheatears.

Most birdwatchers come in late spring or early summer for the breeding birds. By this time Whinchats will have returned to their territories: follow the footslopes, where bracken thrives on the deeper soils, and look for a bird perched on a wall, thistle or tall umbellifer. Young conifer plantations are another likely place. Although there are up to 100 pairs, Twite are more difficult to find. They return late, sometimes not until June, and can be very unobtrusive during the breeding season, as they mix with Linnets and feed in hayfields off the moor. Occasionally they can be seen feeding on roadside weeds and, in late summer, when they flock together, small parties can be seen perched on telegraph wires or feeding on thistles or ragwort.

Moorland is well known as the breeding ground of waders and the North Staffordshire Moors are no exception. In addition to the Curlew already mentioned, the mosses and damp rushy pastures hold good populations of Lapwing and Snipe, the latter at a high density. Each advertises its presence through aerial display flights. On the higher, remoter moors a few pairs of Golden Plover nest each year and very occasionally Dunlin may be seen in display flight, although they do not breed. Of the passage waders, parties of Whimbrel may fly over in May and August, just occasionally pausing to rest awhile.

Raptors are another moorland speciality. Kestrels are always around, but Hen Harrier, Merlin and Hobby also breed, though not necessarily annually. Hen Harriers are very scarce, but might appear anywhere, any time. If found, their languid flight and systematic quartering of an area usually afford good views. Conversely, Merlins seldom offer more than a tantalising glimpse as they dash low and twisting across the heather. Their arrival and departure usually coincides with that of the Meadow Pipits on which they prey and it is probable that a pair or two nest most years. Likewise little is seen of a Hobby when it is in fast aerial pursuit of prey, but on a warm summer's day one might be seen in more leisurely flight as it hunts insects across the heather. Exceptionally a rarity like Montagu's Harrier passes through. Two other moorland predators are the Short-eared and Long-eared Owls. Two or three pairs of Short-eared Owls breed most years and they can regularly be seen quartering rough grassland or young forestry plantations in their quest for voles. Pellets found in young forestry are a sure sign of roosting sites. Long-eared Owls are nocturnal and hence more difficult to find, but the hunger calls of young birds can sometimes be heard coming from a patch of scrub woodland late on a still summer's evening.

Other breeding species are Stonechat, a few pairs of which may nest amongst gorse and scrub; Fieldfare, of which just one pair is usually present; Redpoll, which nest in small birches and sallow thickets; and Teal, a few pairs of which nest in wet ground such as that at Goldsitch Moss. Swarms of hatching insects also attract feeding Swifts and hirundines.

Whilst on the moors do not neglect the scrub, woodland and streams, especially the Blackbrook Valley. Here, among the birch scrub and open woodland that fringes on the moor, Willow Warblers are abundant and there is a good population of Tree Pipits. Within the woods, Chaffinch is the commonest bird, but most species are present in reasonable numbers. Several pairs of the delightful Redstart breed and among the scarcer birds are Sparrowhawk, Pheasant, Woodcock, Tawny Owl, Green Woodpecker, Blackcap, Wood Warbler, Jay, Goldfinch and Redpoll. On the fast, tumbling streams both Dipper and Grey Wagtail are frequently seen, while some 30 pairs of Jackdaws nest on a cliff overlooking the River Dane.

Once they have reared their young, the small birds usually leave their moorland home and move to lower ground. Those that remain tend to flock around a good food source. Ring Ouzels, for example, seek out rowan berries and from late September onwards hordes of Redwing and Fieldfare move into the area to feast on berries. From time-to-time late autumn also brings an exciting vagrant like a Snow Bunting, Black Redstart, Rough-legged Buzzard or wandering Peregrine.

## Timing

The quiet of early morning is always a good time for finding birds. Songsters will be re-establishing their territories in spring or resuming the endless routine of feeding their young in summer. Raptors may be more active from mid-morning onwards, however, once the ground has warmed up and insects are on the wing. Overflying migrants are often brought down by heavy rain or thunderstorms, and in winter they may pass through in large numbers in advance of hard weather. Black Grouse are most active very early in the morning and, in spring, again in the late evening. The moors are becoming increasingly popular for recreation. Rock climbers regularly use Ramshaw Rocks and the Roaches, hang gliders use Hen Cloud and to the east of the A53 the military has a large firing range. (Watch out for red flags which give warning of firing.) The army also uses the area around Swallow Moss for exercises.

## Access

The moors are best approached from the A53 Leek to Buxton road. For the Roaches turn westwards off the A53 at Upper Hulme, 3 miles (5 km) to the north of Leek. After 1 mile (1.5 km) park by the roadside and explore on foot. One suggestion is to take the path that skirts the southern side of Hen Cloud, follow this round to the summit and then head towards the Roaches. From here there is a good expanse of heather moor to the right and superb views beyond the old pinewood on the left. Continue along the Roaches for 2 miles (3 km) into Roach End. From Roach End take the rough track that descends into the Blackbrook Valley. For woodland species follow the path on the left just before the renovated farmhouse and continue into Back Forest. Wander along the tracks in the wood and down to the stream and the River Dane. Return via the top edge of the wood to check for birds feeding on bilberries or heather.

For moorland species stay on the main track from Roach End, past the renovated farmhouse and down into Blackbrook Valley. Look out for Ring Ouzels across the valley. Cross the stream by the footbridge and climb up the hillside opposite to the top of Gradbach Hill, then turn right along the path to regain the road at Goldsitch Moss, where a check should be made for waders. This walk could be made in reverse, beginning at Goldsitch Moss, where cars can be parked by the roadside, or alternatively shortened by beginning at Roach End. There is not much space for parking at Roach End, however, so this can be difficult at busy times.

Another good area for typical moorland birds is Oliver Hill, which is conveniently crossed by a footpath from Flash to Knotbury. This is a good spot for Golden Plover and Twite and affords excellent views to the south.

It is difficult to be precise about the whereabouts of moorland birds as so much depends on the condition of their habitat. This is especially so of heather, which deteriorates both through over-grazing and old age. Most moorland species should be seen in the areas mentioned above, but if not, try the Swallow Moss area. Whilst in this vicinity, stop halfway up the steep hill that approaches Blake Brook from the north-west and scan the fields in the area of Fernyford Farm, as these are sometimes visited by Black Grouse. A telescope will be useful for this.

Finally, the forestry plantations at Gib Torr are always worth a look. They can be checked from the surrounding roads and public footpaths.

## Calendar

*Resident:* Sparrowhawk, Kestrel, Red Grouse, Black Grouse, Tawny Owl, Grey Wagtail, Dipper and corvids.

*March–June:* Teal, Hen Harrier (rare), Merlin (rare), Golden Plover, Lapwing, Dunlin (passage early in period), Snipe, Woodcock, Curlew, Long-eared Owl, Short-eared Owl, Green Woodpecker, Meadow Pipit, Stonechat (scarce), Fieldfare (most leave in April, but a pair may breed) and Redwing (leave in April). Possible passage waders.

From April onwards Hobby (rare), Cuckoo, Tree Pipit, Redstart, Whinchat, Wheatear, Ring Ouzel, common woodland birds including Wood Warbler, Twite (from late May) and Redpoll.

*July–August:* Most summer visitors leave: possible passage waders.

*September–February:* Ring Ouzels leave in September, when first Fieldfares and Redwings arrive. Possible raptor or other rarity, especially in late autumn.

# LONGSDON, RUDYARD & TITTESWORTH

Map 4
OS map 118

## Habitat

These three waters lie near the head of the Churnet Valley, to the west of Leek. None is outstanding on its own, but together they are well worth visiting.

The British Waterways Board (BWB) canal-feeder reservoir at Rudyard covers 175 acres (71 ha). Situated in a deep, natural valley, the lake is narrow but almost 2 miles (3 km) long. Viewed from the A523, high to the east, its setting is magnificent, with steep, wooded slopes and open hillsides above. The lake is quite deep and the banks shelve steeply, so there is little marginal vegetation. At the northern end, though, the water is shallower and the banks more gradual. Here a profusion of water horsetail, yellow flag and reed-grass grades into alder and willow scrub, providing some cover for nesting birds. Low water levels expose mud and this attracts a few migrant waders. Sailing and fishing are both popular and many people visit the lake in summer. There are two small copses on the eastern side, while Rea Cliffe Wood covers much of the western shore. Between them they contain stunted oaks that were formerly coppiced, fine mature sycamores and limes, and a liberal sprinkling of larch and other conifers. Alders thrive down at the water's edge, particularly along the feeder stream below the dam. Bilberry, bracken and wavy hair-grass feature in the acidic ground flora.

Tittesworth Reservoir is the largest of the three, covering 184 acres (74 ha). Principally a water-supply reservoir, administered by the Severn-Trent Water Authority (STWA), it is also used for fishing and there is a small, but popular picnic area. Like Rudyard, it was formed by damming a natural valley, but in this case the shoreline is more irregular and shelves more gradually. The countryside around is more open too, with plenty of upland grazing for winter wildfowl and just one or two conifer plantations. The small, marshy area north of the causeway is good for dabbling duck in winter, while in late summer a low water level exposes some mud for waders. Of the three waters, this is likely to prove the most productive.

Longsdon Mill Pool is no more than a tiny depression in a rough, upland meadow. It, too, is fished and more often than not is devoid of birds. However, the thick growth of sedges, rushes and other marginal plants provides sufficient cover to attract some interesting wildfowl from time-to-time.

## Species

Winter wildfowl, especially swans and geese, and passage waders and terns are the main interest, though all three sites can be erratic.

A few pairs of common waterbirds usually breed, with Great Crested Grebe and Mallard at both reservoirs, Mute Swan at Rudyard and perhaps Canada Goose at Longsdon. By late summer up to 100 Canada Geese may have assembled at Tittesworth, while at Rudyard some 50 or so Grey Herons from the nearby heronry stand around the lake. September often sees the first Wigeon arrive and by the turn of the year a flock of 200 or more grazes the grassy surrounds of Tittesworth. One or two birds are occasionally seen at Rudyard or Longsdon as well. Mallard numbers are often similar, but the birds are more dispersed, with between 150 and 200 at Tittesworth and perhaps 40 to 50 at Rudyard. Numbers some years, though, are considerably lower. At Tittesworth, much depends on the water level in the marshy area favoured by dabbling duck. Teal are very unpredictable. Typically there are less than 20 anywhere, but periodic influxes in November may bring over 100 to either of the two reservoirs, while 60 have been seen in mid-winter at Longsdon. Of the remaining dabbling duck, Shoveler are irregular, one or two Gadwall and Pintail appear most winters, and Shelduck regularly pass through in late autumn and early spring.

Diving birds are equally variable, but small numbers of Pochard and Tufted Duck are usually present. Rudyard seldom has more than a dozen of each, but at Tittesworth over 50 may be present, with exceptionally well over 200. A few Goldeneye and Goosander are also regularly noted at both waters. Up to 100 Coot and a few Great Crested Grebes also frequent Tittesworth, Cormorants and Ruddy Duck are noted on occasions and Common Scoter are sometimes seen on passage. Among the less likely visitors, Great Northern Diver, Scaup and Red-breasted Merganser have all been seen at Tittesworth.

Above all, this area is known for its wintering swans. Up to a dozen Whooper Swans have been regular visitors since 1976. They normally arrive in November or December and stay until March or early April. During this time they wander between the three waters, sometimes splitting up, but frequently returning to the small Longsdon Mill Pool. Bewick's Swans are less reliable, but small herds up to six have been recorded most years at some time between October and March. Again Longsdon has been visited by this species. Indeed, regular watching here has also produced an autumn Bittern and a fine drake Garganey in spring, though these must be regarded as unusual.

Geese are more enigmatic. Canada Geese are resident, with winter flocks of up to 100 at Tittesworth and 20 or so at Rudyard, and a few feral Greylags are often at the former locality. From time-to-time they are joined by one or two Pink-feet or a Barnacle Goose. In the past these too were believed to be feral, but in recent years skeins of 150 Pink-footed Geese have twice been seen flying down the Churnet Valley, accompanied on one occasion by two Barnacle Geese. In addition, a flock of 60 Pink-feet remained at Tittesworth throughout February 1982 and one or two birds have been seen most winters since. The large flocks probably arise from hard weather movements of wild birds from the Lancashire mosses, but the origins of the smaller parties remain open to question. Brent Geese, too, have been recent autumn visitors.

Waders are of great interest. Passage begins in March, when up to 100 and sometimes twice as many Curlew may visit Tittesworth *en route* to their moorland breeding haunts. In the evening many leave for Longsdon, where there is a huge roost of over 300 every year (and once, twice as many). Little Ringed Plover and Common Sandpiper then return in April; in a good year half-a-dozen pairs of each attempt to breed around the reservoirs with varying success. Just occasionally a pair of Redshank nest as well, while two or three Snipe regularly display at Tittesworth and over the wet meadows at Longsdon. Breeding species apart, spring wader passage is often poor, though Oystercatcher and Dunlin are regularly noted and Ringed Plover, Sanderling, Bar-tailed

Godwit and Greenshank are occasionally seen. Scarce spring visitors have included Grey Plover, Black-tailed Godwit, Whimbrel, Wood Sandpiper and Turnstone. Tittesworth is generally the better reservoir for passage waders at this season.

If the breeding season has been good there may be up to two dozen Little Ringed Plovers and Common Sandpipers by July. Soon they are joined by the first returning passage birds. Green Sandpipers show early, with four or five a typical number and perhaps a Wood Sandpiper with them. August then brings a few Ringed Plover and Dunlin, one or two Whimbrel and Ruff, and half-a-dozen Greenshank. An Oystercatcher is seen most years and less regularly a Knot, Black-tailed Godwit, Spotted Redshank, Little Stint or even a Curlew Sandpiper might appear. Totally unexpected, though, was an adult White-rumped Sandpiper at Tittesworth. After the breeding season many waders move down from the moors to the reservoirs again, and large gatherings of Curlew and Snipe occur in late summer. Curlew are seldom as numerous as in spring, but over 100 Snipe may be probing for food at either Rudyard or Tittesworth by late August or September. Many may stay well into a frost-free winter, being joined perhaps by a Jack Snipe in November or December, and if very mild even a Redshank or Common Sandpiper may tarry. Overall, Rudyard is perhaps marginally the better for autumn passage.

Gulls and terns are never numerous, but one or two Common or Arctic Terns pass through in spring or autumn and a Little or Black Tern is occasionally seen. With the terns may come a couple of Little Gulls or a storm-driven Kittiwake. In winter, the reservoirs, particularly Tittesworth, hold a few gulls and a Water Rail might secrete itself in the marginal vegetation. Kingfishers often breed in the vicinity of the reservoirs and might be seen any time; Grey and Pied Wagtails frequent the streams below the dams, and one or two pairs of Sedge Warbler breed.

Away from the water there is a good range of common woodland birds, especially around Rudyard. Here the residents include all three species of woodpecker and Nuthatch, whilst summer brings Yellow Wagtail, Redstart and many warblers including Grasshopper and Wood. Sparrowhawk and Woodcock also breed in the conifer plantations, but the latter is more numerous in winter. This is also the season when Siskin congregate in the alders, especially along the canal-feeder arm below the dam at Rudyard, and Tree Sparrows, Chaffinches, Linnets and other finches flock to the surrounding pastures, where at dusk they might fall prey to a hunting Barn Owl. Exceptionally, a few Bramblings are present or a Snow Bunting is seen around one of the reservoirs in November. Although Sparrowhawk and Kestrel are the commonest raptors, passage may bring a Buzzard, Osprey or Peregrine to enliven the day.

## Timing

All three sites are fished, the picnic areas are extremely popular in summer and there is sailing too. Early visits are therefore advisable, particularly in spring and summer. In winter, weather is more important than disturbance. Freezing conditions or heavy snowfalls further north may well bring fresh wildfowl, possibly including geese.

## Access

Rudyard is reached from the A523 Leek to Macclesfield road. About 1½ miles (2.5 km) north of Leek turn westwards onto the B5331 and follow this into Rudyard. Turn right at the mini-roundabout in the village and then fork immediately right into Lake Road. There is a car park on the right-hand side immediately beyond the Hotel Rudyard. From here a path runs right round the reservoir.

Alternatively continue along the A523 for a further 3 miles (5 km), then turn westwards at Ryecroft Gate. In a few yards, immediately over the old railway bridge, turn left and follow a track alongside the old railway line. At the fork, either park to explore the western shore on foot, or bear to the left and drive under the bridge and along the eastern shore where there are parking areas.

Tittesworth is best approached from the A53 Leek to Buxton road. Some 2½ miles (4 km) north of Leek turn westwards at the Three Horseshoes Inn towards Meerbrook. After about 1 mile (1.6 km) the road crosses the reservoir on a causeway. Either park here, or (in summer) at the picnic site on the left.

To view Longsdon Mill Pool, turn northwards off the A53 Stoke to Leek road at the cross roads ¼ mile (0.4 km) to the east of Longsdon Church. After ½ mile (0.8 km) turn right into another lane and the pool can be viewed from the road about ½ mile (0.8 km) further on.

## Calendar

*Resident:* Great Crested Grebe, Grey Heron, Mute Swan, Canada Geese, Mallard, Sparrowhawk, Kestrel, Snipe, Woodcock, Kingfisher, all three woodpeckers, Grey Wagtail and common passerines.

*April–June:* Oystercatcher, Little Ringed Plover, Ringed Plover, Sanderling (scarce), Dunlin, Bar-tailed Godwit, Curlew, Greenshank (scarce), Common Sandpiper, and perhaps other waders; possible Little Gull or Common, Arctic, Little or Black Tern; Yellow Wagtail, Redstart and warblers including Grasshopper, Sedge and Wood. Maybe a passing raptor.

*July–September:* Oystercatcher, Little Ringed Plover, Ringed Plover, Dunlin, Ruff, Whimbrel, Curlew, Redshank, Greenshank, Green Sandpiper, Common Sandpiper and perhaps Knot, Little Stint, Curlew Sandpiper (rare), Black-tailed Godwit, Spotted Redshank or Wood Sandpiper; possible Little Gull or terns as above. Maybe an Osprey or other passing raptor.

*October–March:* Cormorant (scarce), Bewick's Swan (scarce), Whooper Swan, Pink-footed Goose (scarce), Shelduck (early and late), Wigeon, Gadwall (scarce), Teal, Pintail (rare), Shoveler (scarce), Pochard, Tufted Duck, Goldeneye, Goosander, Ruddy Duck, Water Rail, Coot, Jack Snipe (scarce), Curlew (March), gulls, Barn Owl (scarce) and flocks of thrushes and finches including Fieldfare, Brambling (scarce) and Siskin. Perhaps diver, geese, sea-duck, raptor or Snow Bunting.

## Habitat

The Churnet rises on the moors of north Staffordshire and flows 25 miles (40 km) to join the Dove near Uttoxeter. Gouged out by glacial meltwaters, its valley is 330 ft (100 m) deep and gorge-like in places. This, combined with the Germanic local architecture, has led to its being called 'the Staffordshire Rhineland'. The main river is sluggish, with broad meanders in the lower reaches, but the tributaries are faster flowing, with shallows, shingle spits and boulder-strewn reaches.

The narrow valley is lined with small, lush, grazing meadows. The recently restored Caldon Canal and a partially restored old railway follow the course of the valley. The land between these and the river is often marshy and liable to flood. Thistles grow in meadows enclosed by hawthorn hedges, alders flourish along the canal and river banks, and there is a profusion of meadowsweet and great hairy willowherb. Below Cheddleton arrowhead grows in the canal, while reed-grass provides cover for nesting birds in many places.

Mixed woodland clothes the steep hillsides, forming an almost continuous canopy from Longsdon down to Alton. Much of the original ancient woodland has been felled to be reclaimed for agriculture or replanted with alien larch, pine, spruce or cypress. Often there is an amenity fringe of oak, ash or sycamore and an understorey of rhododendron that provides winter roost sites. Of the conifer plantations, that of the Forestry Commission (FC) at Consall Wood is one of the largest and most varied.

Hardwoods survive in good variety too. The National Trust (NT) has a nature reserve at Hawksmoor which has stunted oaks and birches scattered over a hillside of bracken, heather and bilberry. At Consall Forge both river and canal flow through a remote and very beautiful bowl of semi-natural ash woodland. Equally beautiful is the area around Oakamoor, where private and Forestry Commission woodlands blend into an exquisite mixture of hard and softwoods.

For birds, though, the sessile oakwoods are most important and many are now protected by the RSPB and other conservation bodies. Coombes Valley is typical of most and the easiest to visit. Here, in addition to the oaks, are ash, birch, holly, rowan and wych elm above a shrub layer rich in bird cherry, blackthorn, hazel, guelder rose and other food plants. There are also hillsides of bracken and heather, meadows by the stream and a small pool overlooked by a hide. A second hide in the oaks that overhang the stream gives excellent views of many woodland species.

## Species

Woodland and waterside birds are the main attraction of this picturesque valley, particularly in spring and summer.

To see woodland residents like the three woodpeckers, Nuthatch, Treecreeper, thrushes, tits and finches an early spring visit is recommended. Even on chilly mornings birds are then beginning to sing and once located they are easily observed in the bare trees. This is also a good time to look for Goldcrests. Coal Tits and perhaps Redpoll or Siskin in the conifer plantations, where the Siskin favour pine and spruce.

For migrant songbirds a later visit will prove more rewarding, as many do not return to these upland valley woods until late spring. From mid-May onwards the woods resound to a chorus from warblers such as Blackcap, Garden Warbler, Whitethroat, Chiffchaff and Willow Warbler as well as the resident birds. Amongst scattered trees on bracken-covered hillsides, Tree Pipits indulge in their characteristic parachute display flights. Higher up the slopes, Whitethroats, Linnets and Yellowhammers nest in gorse and thorn scrub. In the background, Cuckoos call and Skylarks sing from high above their meadow nests. A few pairs of Lapwing and maybe a pair of Snipe or Yellow Wagtail breed in one or two of the damper meadows.

Three migrant songbirds are especially characteristic of the Churnet Valley — each with its distinct habitat requirement. Wherever there is a closed canopy and sparse shrub layer, the sibilating song of Wood Warblers is commonplace. They occur usually under stands of beech or in hanging oakwoods. Redstarts, however, prefer gaps in the canopy and a well-developed shrub layer. There is a particularly good population of over 30 pairs at Coombes Valley. This is also the best place to see the third speciality — the Pied Flycatcher. They like steeply sloping oakwoods overhanging a stream, and commonly breed in nest-boxes. At Coombes Valley some ten pairs nest. They are best watched from the tree-hide, where the smart black-and-white males flit from branch to branch.

*Redstarts are particularly numerous in the Churnet woodlands*

Sparrowhawks are the commonest raptor and often soar or spiral on the thermals that rise above the valley on warm, sunny days. Occasionally they pursue and mob a larger raptor, such as a stray Buzzard. Rooks, Carrion Crows and Jackdaws are all numerous, while the Magpie and Jay populations are far healthier than is good for the smaller song birds.

A quiet evening stroll might be rewarded with a Woodcock roding over the birches and bracken, a Little Owl hunting in the twilight, or Tawny Owls calling from a distant wood. Exceptionally a Long-eared

Owl is encountered. One or two pairs are regularly present, but not easily located. They might be detected by their call, which is more of a moan than a hoot.

By high summer, when young birds are being fed by their parents, the woods are unexpectedly quiet. Come late July or early August, family parties of summer visitors slip quietly away and many resident adult birds go into moult.

Activity revives again in autumn, by which time the valley woods are looking their best in a mantle of browns, golds, reds and yellows. From early October onwards hordes of migrant Redwing and Fieldfare avidly feed on the abundance of hips, haws and other berries in wood and hedgerow. A good year brings several thousand of each and large roosts gather of an evening, especially in woods with an understorey of rhododendron. With them come increased numbers of Blackbirds, Song and Mistle Thrushes and perhaps a late Ring Ouzel. If there is a good crop of beechmast, a few Brambling may join the usual flocks of Chaffinches and foraging tits. Chaffinches also forage in the meadows and around stubbles and ploughed fields, where they are often joined by Tree Sparrows, buntings and other finches. In one of the wet, marshy corners you might be lucky enough to glimpse a Water Rail or catch sight of a Barn Owl at dusk. Small flocks of Goldfinch, Redpoll and Siskin regularly come to feed among the alders each autumn. Most birds arriving in these autumn influxes remain until either their food supply is exhausted or severe weather forces them to seek a milder clime.

Waterside birds can be seen throughout the year. The streams that twist and fall down the steep slopes into the Churnet are a favoured haunt of Dippers and Grey Wagtails, but Kingfishers prefer the stiller reaches of the main river. All three species can be seen from the pool hide at Coombes Valley. Mallard and Moorhen raise broods along the canal, where Swallows and House Martins regularly feed and Grey Herons come from the nearby Longsdon heronry to fish.

On the whole the valley would repay more regular watching. It is hardly the place to expect rarities, but a Black Stork has recently visited twice in three years, so who knows what to expect?

## Timing

Early morning is invariably most productive for songbirds, especially in spring and early summer. By high summer all woodland species have become more secretive. A late evening visit is necessary for roding Woodcock and owls. Timing matters less in winter, when disturbance is reduced and most birds are active throughout the day. But remember, little will be seen in wet or windy weather and prolonged cold spells may force all but the hardiest species to forsake the area altogether. Waterbirds can be seen at most times of the day.

## Access

All the characteristic birds can be seen at the RSPBs Coombes Valley reserve. Leave the A520 Leek to Stone road eastwards on one of the minor roads leading to Basford Bridge (Cheddleton Station). Continue past the station for ¾ mile (1.2 km) and then turn left into a narrow lane that climbs steeply. The reserve is on the right-hand side a little over 1½ miles (2.5 km) along this lane. Park in the car park and contact the warden in the Information Centre for details on access and what's about.

The best points to explore the Churnet Valley are as follows:

*From Cheddleton Station* Follow the Staffordshire Way southwards along the towpath of the Caldon Canal. Remain on the towpath to Froghall, a distance of about 4½ miles (7.3 km). Check the small meadows *en route*. The last 2 miles (3 km) of this stretch pass through some of the finest scenery in the valley, and both woodland and waterside species can be seen. Alternatively the walk can be approached from Froghall, on the A52 Stoke to Ashbourne road, where cars can be parked in the picnic area, provided by the Staffordshire County Council.

*From Alton to Oakamoor* For better car access take the narrow lane which runs along the valley. Alton Village is reached from the B5032 Rocester to Cheadle road. (Do *not* follow the signs to Alton Towers.) Turn left just before the bridge in the village into the little lane that follows the river. In about 1 mile (1.5 km) park by the roadside and follow the Staffordshire Way into Dimmingsdale — a small valley to the left of the road. After ⅓ mile (0.5 km) one path follows Dimmingsdale to the left, while the Staffordshire Way continues straight ahead into Ousal Dale. Follow either path to a little lane, then return by the other route to complete a good circular walk.

Continue by car to Oakamoor and turn left onto the B5417 towards Cheadle. In a little over 1 mile (2 km) park on the right-hand side and explore the open hillsides and woods of the NT nature reserve at Hawksmoor Wood. (A guide book to the nature trails on the reserve is available in Oakamoor.) Staffordshire County Council has another picnic area on the B5417 at Oakamoor. From here there is a pleasant, easy walk of 2 miles (3 km) to Alton and 4½ miles (7.5 km) to Denstone along the old railway line.

Finally, the really energetic can follow the Staffordshire Way for nearly 20 miles (32 km) from Rudyard in the north to Rocester in the south.

## Calendar

*Resident:* Grey Heron, Mallard, Sparrowhawk, Kestrel, Moorhen, Lapwing, Snipe (scarce), Woodcock, Barn Owl (scarce), Little Owl, Tawny Owl, Long-eared Owl, Kingfisher, all three woodpeckers, Grey Wagtail, Dipper, Nuthatch, Treecreeper, Redpoll and other common woodland birds.

*April–June:* Cuckoo, Swallow, House Martin, Tree Pipit, Yellow Wagtail, Redstart, woodland and scrub warblers including Wood Warbler, Spotted Flycatcher and Pied Flycatcher.

*July–September:* Generally quiet as summer visitors depart.

*October–March:* Water Rail (scarce), large roosts and flocks of thrushes including Fieldfare and Redwing, flocks of Tree Sparrows, finches and buntings including Brambling and Siskin. Perhaps a late Ring Ouzel in October.

## Habitat

Stoke-on-Trent is undergoing an urban renaissance, exemplified by the 1986 National Garden Festival. No longer a place of grimy collieries, belching chimneys and innumerable bottle kilns, it is fast becoming a greener, more pleasant city through a massive programme of derelict land reclamation. Stoke developed through the gradual coalescence of its 'five towns'. This left pockets of undeveloped land, to which subsidence and dereliction contributed further. Such areas soon became 'wastelands' of willowherb, nettle, thistle and rank grass, with marshy hollows of rush, sedge, reedmace and reed-grass where subsidence caused flooding. Elsewhere, old spoil mounds, devoid of top soil and prone to gully erosion because of their steepness, were slowly recolonised by coarse grassland and scrub. All too often such areas became much abused wastes, serving as playgrounds, exercise areas for dogs and rubbish tips. The derelict land reclamation programme aims to manage and accelerate natural regeneration so as to create more useful and attractive open spaces.

## Species

In some respects industrial dereliction will be missed. Wheatear, Yellow Wagtail, and Meadow and Tree Pipits, for example, are all recorded as having nested on the slag heaps of Etruria and Longton. Perhaps some of these birds can still be found on a remaining tip and only recently one such spot was visited by a migrant Wryneck. Kestrels and occasionally owls still nest in derelict buildings and Meadow Pipits breed in fair numbers among the rank grasslands. Such areas support a few pairs of Grey Partridge and Whinchat as well. In late summer and autumn, thistle-heads and weed-seeds draw parties of Goldfinches and other finches. The marshy ground of subsidence areas provides a safe autumn roost for hirundines and wagtails, and a winter refuge for Mallard, Teal, Snipe and even an occasional Jack Snipe. Where subsidence has been severe enough to create standing water, Great Crested and Little Grebes and Mute Swan may breed and Kingfishers can be seen. The more interesting sites are detailed below.

## WESTPORT LAKE

## Habitat

The principal feature of the 100 acre (40 ha) Westport Lake Park is a shallow 25 acre (10 ha) lake formed by the subsidence of marl workings. The lake is administered by the City of Stoke-on-Trent. There are also three smaller pools, plus a marshy area of reed-grass and water horsetail which has been designated a nature reserve. Flanking this reserve is a backdrop of alder and willow. Otherwise this is essentially a new landscape of earth mounds and tree planting which is not yet mature enough to really attract birds.

## Species

Westport Lake is probably the best site in the Potteries for wintering wildfowl and gulls, and for passage terns and waders. Numbers are not spectacular, but the variety is good. Typically some 20 to 30 Mute Swans are present throughout the year, being joined in winter by up to 200 Tufted Duck, 100 or so Pochard, 10 Great Crested Grebes, half-a-dozen Goldeneye, 200 Coot and an occasional Goosander. The lake is less favoured by dabbling duck, but up to 50 Mallard and one or two Wigeon can usually be seen, while Teal, Gadwall, Shoveler and Pintail all make irregular visits. One or two Ruddy Ducks are often present as well and breeding has occurred at least once. Great Crested Grebes also breed and Mallard do so on occasions. One or two Cormorants and Shelduck are noted most years, while recent appearances of scarcer species like Great Northern Diver, Shag, Bewick's and Whooper Swans, Garganey, Scaup and Common Scoter show what regular watching can reveal.

Most of the gulls from the Potteries roost on meres and lakes in Cheshire, so numbers at Westport are seldom large. Normally there are between 8,000 and 10,000 Black-headed Gulls, around 500 each of Lesser Black-backed and Herring, some 50 Common and one or two Great Black-backed. Despite these small numbers, either an Iceland or a Glaucous Gull is seen most winters, and Kittiwakes are almost annual visitors, often in spring.

Spring, and more especially autumn, bring a few passage waders and terns. Among the waders, Common Sandpiper is most regular, but Dunlin, Sanderling and Little Ringed Plover have all appeared at both seasons, with the latter attempting to breed. Ringed Plovers are most likely to pass through in spring, but Greenshanks are a late-summer visitor. Oystercatcher, Redshank and Turnstone are among the commoner species to have visited, while rarities have included a party of six Avocets, Kentish Plover and Grey Phalarope. Small parties of Common and Arctic Terns also pass through in spring and autumn. A solitary Little Tern in June is by no means exceptional, this seemingly being a regular staging post, but Black Terns are less predictable.

Of the passerines, a White Wagtail may enliven spring, while a few Sedge and Reed Warblers add their song to the long summer days. Return passage sees a good variety of warblers moving through in August, followed by Swallows and then Meadow Pipits. Late autumn may also bring Rock Pipit, Water Rail or passage Merlin. Finally, as winter sets in, the alders are visited by feeding flocks of Redpoll, Siskin and Goldfinches, and up to 50 Magpies come in to roost.

## Timing

As the main venue for water sports in Stoke-on-Trent, Westport is popular at all times and very crowded on summer weekends. It caters for sailing, fishing, swimming and casual recreation. Fortunately sailing is prohibited from November to March, so this enables duck numbers to build up. Early morning is likely to be the quietest time, but late afternoon can be good for gulls.

## Access

Westport is situated a little under 1 mile (1.5 km) due west of Burslem town centre, off the A527 Newcastle to Congleton road. From the A500, which is the main Potteries spine road, turn eastwards onto Porthill Road (A527), cross over the railway and canal bridges and continue into

Davenport Street. Almost immediately there is a left turn back over the canal and into Westport Lake Park, where car parking overlooks the lake. Follow the established paths to explore the park.

## Calendar

*Resident:* Great Crested Grebe, Mute Swan, Mallard, Coot and common passerines.

*April–June:* Shelduck, Oystercatcher, Little Ringed Plover, Ringed Plover, Sanderling (scarce), Dunlin, Redshank, Common Sandpiper, Turnstone (scarce), Kittiwake (scarce), terns including perhaps Little and Black, White Wagtail (scarce), Sedge Warbler and Reed Warbler. Possible passage rarity.

*July–September:* Little Ringed Plover, Sanderling (rare), Dunlin, Greenshank, Common Sandpiper, terns including perhaps Black, Swallow, and passage warblers.

*October–March:* Cormorant (scarce), Shelduck, Wigeon, Gadwall, Teal, Pintail (rare), Shoveler, Pochard, Tufted Duck, Goldeneye, Goosander (scarce), Ruddy Duck, Water Rail (autumn: scarce), Jack Snipe (autumn: rare), Snipe, gulls including occasional Iceland and Glaucous, (after December), Meadow Pipit (October), Rock Pipit (rare) and flocks of passerines including Goldfinch, Siskin and Redpoll. Perhaps a passage raptor, one or two scarce wildfowl or a vagrant seabird.

## FORD GREEN

### Habitat
This is a small area of marshy ground and open water in a shallow valley alongside the B5051. Surrounded by an expanse of grassland, the marsh itself is very wet and its stands of reedmace and reed-grass are relatively safe from disturbance. In winter the open water is large enough to hold a few wildfowl.

### Species
The main interest at this small site is the autumn hirundine and wagtail roost. Small numbers of Swallows begin to roost as early as August and by September some 12,000 to 15,000 are regularly present, with up to twice that number on occasions. At this time of year 200 or 300 Yellow Wagtails also roost. Pied Wagtails come later, peaking at around 300 in October, when a solitary Water Rail might also be seen skulking amongst the reedmace. There is also an autumn build up of Snipe, with between 50 and 100 during November and December, when the odd Jack Snipe may be found.

During a wet winter, floodwater will attract a few Mallard and up to 50 Teal, and even a family party of Bewick's Swans might pay a brief visit. As the water level falls, flocks of finches, especially Linnets, gather to feed on exposed seeds.

Spring sees a return passage of Snipe and Jack Snipe, the latter usually in April or early May. It also brings an occasional migrant wader

*Roosting Swallows are a feature of Ford Green*

like Little Ringed Plover. Wagtails, too, move through again and with the Pieds there is always the chance of the paler, grey-rumped White.

The emergent vegetation provides just sufficient cover for one or two pairs of Little Grebe and Tufted Duck to nest and in a good year a Grasshopper Warbler may be heard reeling in rough areas. The grassland also supports many small mammals and these attract feeding Kestrels or exceptionally even a Short-eared Owl.

## Timing

The area is well used by the local inhabitants at all times. Early mornings, before there has been any disturbance, are likely to be best for Snipe and Jack Snipe, but dusk is best for roosting birds.

## Access

Ford Green lies 1½ miles (2.5 km) north-east of Burslem town centre, immediately north of the B5051. Leave Burslem along Moorland Road (B5051) and continue into Ford Green Road. Park near the bottom of the hill and the marsh is on the left-hand side. There is an obvious path around the perimeter

## Calendar

*Resident:* Little Grebe, Tufted Duck and Kestrel.

*March–June:* Jack Snipe (scarce: not June), Snipe and wagtails including perhaps White. Maybe a passage wader (such as Little Ringed Plover) or Grasshopper Warbler.

*August–September:* Roosting Swallows and Yellow Wagtails.

*October–March:* Mallard, Teal, Water Rail (scarce), Jack Snipe (scarce), Snipe, Pied Wagtail (roost in October) and flocks of finches especially Linnet.

29

## Park Hall Country Park

### Habitat

Between the housing estates of Bentilee and Weston Coyney, the hilly 333 acre (135 ha) Park Hall site represents one of the more ambitious reclamation schemes in Stoke-on-Trent. Formerly a deer park, the area was despoiled firstly by coal mining, then by sand and gravel extraction and finally by waste disposal. The result was a desolate and dangerous wasteland of pit shafts, sludge tips, settling lagoons and unfenced canyons. Now it has been reclaimed by Staffordshire County Council and the City of Stoke-on Trent into a Country Park, with a mosaic of heathland, woodland, secluded valleys and pools.

The park sits astride a sandstone ridge and parts of it still support the indigenous heathland flora, with small patches of heather, gorse and scrub. Evocative names like Khyber Pass and Old Man of Hoy have been given to the distinctive canyons and sandstone outcrops from which there are panoramic views. There is also an old pinewood, extensive new grasslands sewn with wildflower seed, and nearly 100,000 newly-planted trees. In the south-western corner, the Lady's Corner Pools provide a home for wildfowl and migrant waders.

### Species

In summer Linnets and Yellowhammers, which nest in gorse and scrub, add their voices to those of Skylarks and Meadow Pipits, which breed commonly on the heath and grasslands. More sparingly, one or two pairs of Whinchat breed and the brightly plumaged cock might be seen perched on low vegetation. With luck, a Grasshopper Warbler might also be heard reeling from a patch of bramble or other suitable cover.

Spring and autumn bring passage Wheatears and just occasionally a migrant Stonechat may appear. The heath is good for raptors too, with Kestrel and Sparrowhawk regularly seen throughout the year and always a chance of a Short-eared Owl hunting for voles in autumn, winter or early spring. Little Owls are often in the vicinity as well, and Barn Owls have been seen in the past.

In the pinewoods, several pairs of Redpoll breed and the high-pitched calls of Goldcrests can often be heard. They are especially numerous in winter, when they are joined by foraging flocks of tits and finches. With luck you might even encounter a Long-eared Owl.

The reedy pools near Lady's Corner provide just sufficient cover for Little Grebe, Mallard and Tufted Duck to breed. More interestingly they harbour a few Snipe and very occasionally a passage wader like Dunlin, Golden Plover, Redshank or Greenshank.

Overall this is a developing site with considerable potential. Not a great deal is known about its birdlife at present, but it would certainly repay more regular watching.

### Timing

An early morning visit, when birds are singing and not too many people have arrived, is recommended in summer. Otherwise not really critical, though parts are very exposed so little may be seen in windy weather.

### Access

The main entrance to the country park is north of the B5040. About ¾ mile (1 km) west of the traffic lights in Weston Coyney turn right into

Hulme Road and follow the signs to the Visitor Centre on the right-hand side, where there is plenty of parking. The whole park can be explored from here. For Lady's Corner pools and marsh use the car park on the right-hand side of Hulme Road before the Visitor Centre.

For the conifer woodland, follow the A520 Stone to Leek road for ½ mile (0.8 km) north of the traffic lights in Weston Coyney. Turn left into the country park at the end of the houses and park.

A network of footpaths provides circular walks from any of the car parks. The Visitor Centre is open Sunday afternoons in the summer.

## Calendar

*Resident:* Little Grebe, Mallard, Tufted Duck, Sparrowhawk, Kestrel, Little Owl, Long-eared Owl, Skylark, Meadow Pipit, Goldcrest, Linnet, Redpoll, Yellowhammer and other common passerines.

*April–June:* Whinchat, Stonechat, Wheatear and perhaps Grasshopper Warbler. Occasionally a passage wader.

*July–September:* Passage Whinchat, Stonechat, Wheatear and occasional wader.

*October–March:* Snipe and foraging flocks of tits and finches. Perhaps a Short-eared Owl.

## TRENTHAM PARK

### Habitat

Though not a reclamation site, no account of the Potteries would be complete without mention of Trentham Park. Here splendid formal gardens gradually blend into a more natural landscape of broad-leaved woodland, parkland and scrub around a large lake.

The lake is used for water sports, a miniature railway skirts one shore and there is a caravan site and golf course. Despite these intrusions, many birds still manage to find a quiet sanctuary in which to nest, feed or roost. The three wooded islands, with overhanging rhododendrons, shelter a few wildfowl, while many woodland species wander into the gardens from the private wooded hillside above. There is also an alder-lined stream to the north and a smaller pool, known as Black Lake, alongside the M6 motorway. The site has had a chequered history and is currently in the ownership of British Coal.

### Species

Up to a dozen pairs of Great Crested Grebe and a few pairs of Mallard breed most years beneath the overhanging vegetation around the main lake, while Little Grebes raise their young on the smaller Black Lake. Grey Herons visit at all seasons and occasionally attempt to nest on one of the islands, but usually without success. Kingfishers sometimes nest in the nearby river bank, but come into the park to fish. The small stream at the northern end of the park is a good place to see them. Dippers, too, have been seen here once or twice in spring. Check the bridges around the lake as they are favourite spots for Grey Wagtails.

The surrounding broad-leaved woodland supports a good range of breeding birds. All three woodpeckers are present, ensuring plenty of holes for tits, Nuthatches and a few migrant Redstarts, while Stock Doves and Tawny Owls inhabit the larger, natural holes. Summer brings Blackcap, Garden Warbler, a few Wood Warblers, Spotted Flycatcher and maybe a passage Pied Flycatcher. In the conifers Goldcrests and Coal Tits are common and there is a good population of Redpoll, while around the woodland fringe Woodcock rode and one or two pairs of Tree Pipit breed. Treecreepers are quite numerous and in winter roost behind the bark of giant Wellingtonias.

Spring and autumn usually bring one or two passage terns, mostly Common but sometimes Arctic or Black, and an occasional Common Sandpiper or Redshank.

In winter there are small numbers of wildfowl, with around 200 Mallard and Coot, 20 or so Tufted Duck and up to a dozen Mute Swans, Canada Geese and Pochard. A handful of Teal are often present as well, while Wigeon, Gadwall and Shoveler are occasionally seen. A few wildfowl also frequent Black Lake. In the woods at this time, one or two Brambling and maybe an over-wintering Blackcap sometimes join the resident species. Up to 100 Mistle Thrushes feast on the yew berries in autumn, while on winter evenings flocks of finches, Redwings, Song Thrushes and Blackbirds come into the dense rhododendron cover to roost. The alders along the northern stream usually produce a few Siskin.

A Sparrowhawk might be seen circling above the woods at any time and Buzzards make periodic visits usually in spring or autumn. Vagrants are few, but Kittiwakes and an Arctic Skua have been seen in recent years and two Ospreys once graced the lake.

## Timing

The gardens are open throughout the year, but get very crowded in summer. For woodland songsters, therefore, an early morning visit is best. Most water sports take place at weekends.

## Access

The gardens (where a charge may be made) are on the west side of the A34 Newcastle to Stone road, 3½ miles (6 km) south of Newcastle town centre. The entrance is off a roundabout at the junction with the A5035 to Longton. There is ample car parking within the gardens and obvious paths to the lake.

The parkland can be reached from the A519 Stoke to Newport road, south of the junction with the M6 motorway (Junction 15). Two footbridges lead directly off this main road across the motorway and into the park, while a third footbridge leads off a lane on the left that goes towards Tittensor. Park along the verges and follow public footpaths across the park.

## Calendar

*Resident:* Little Grebe, Great Crested Grebe, Grey Heron, Mallard, Sparrowhawk, Woodcock, Stock Dove, Tawny Owl, Kingfisher, all three woodpeckers, Grey Wagtail and common woodland passerines including Goldcrest, Coal Tit, Nuthatch, Treecreeper and Redpoll.

*April–June:* Tree Pipit, Redstart, warblers including Blackcap, Garden Warbler and Wood Warbler, and Spotted Flycatcher. Passage terns and occasional Common Sandpiper or Redshank. Perhaps Buzzard, Dipper or Pied Flycatcher.

*July–September:* Passage terns and occasional Common Sandpiper or Redshank. Maybe a migrant Buzzard or Osprey.

*October–March:* Mute Swan, Canada Geese, Wigeon (scarce), Gadwall (scarce), Teal, Shoveler (scarce), Pochard, Tufted Duck, Mistle Thrush (yews in late autumn), roosting thrushes including Redwing, and finch flocks including Brambling (scarce) and Siskin. Maybe a wintering Blackcap.

# BLITHFIELD RESERVOIR

Map 8
OS map 128

## Habitat

Blithfield is a water-supply reservoir administered by the South Staffordshire Waterworks Company (SSWC). It has the configuration of a letter Y and lies in open countryside 4 miles (6 km) north of Rugeley. It is divided into two parts by a causeway carrying the B5013 Rugeley to Uttoxeter road. With an area of 790 acres (320 ha) and a perimeter of 9 miles (14 km), Blithfield is the largest water in the region. It is also arguably the best ornithologically, and is certainly the best for waders.

The deepest water is south of the causeway, behind the impounding dam at the southernmost end. This dam apart, the shoreline is largely natural, with gently shelving, grassy slopes grazed by wildfowl. The shoreline is indented and many bays have rushy edges which provide cover for dabbling duck. Low water levels expose muddy margins that attract a few migrant waders. Sailing and fly-fishing cause disturbance on this deeper, southern section. Nevertheless birds still make good use of it and it remains the favoured area for Wigeon.

North of the causeway the water divides into the two arms of the letter Y, fed respectively by the River Blithe (west) and Tad Brook (east). At this end of the reservoir there is fly-fishing, both from bank and boat, but no sailing, so there is less disturbance. This, coupled with the more varied habitat, means that it usually holds the greater number and variety of birds.

The banks shelve very gently, so even a moderate fall in water level uncovers a good area of mud. After a dry summer, the area of mud exposed at the head of the two arms can be quite extensive, providing by far the most important habitat in the region for autumn waders. Unfortunately such conditions rarely occur in spring. Above these expanses of mud, both arms are flanked by marshy areas of rush and reed-grass. The Blithe is also lined by alders. Dense willow scrub is now encroaching onto the mudflats. Whilst this provides cover and nest-sites for Great Crested Grebes, it is threatening to engulf the wader habitat and is therefore being controlled.

Stansley Wood, on the point between the two arms, is a remnant of mature deciduous woodland. Despite much felling and replanting with conifers, it still holds a good mix of typical woodland birds. On the eastern shore of Tad Bay, the plantations of spruce and larch are ornithologically less interesting.

## Species

There is usually something interesting to be seen throughout the year. Apart from waders, Blithfield is particularly important for its number and variety of wintering wildfowl, ranking amongst the top half-dozen waters in Britain. Over 30 species have been recorded, including the nationally rare Ring-necked Duck and several that are very rare inland.

All the commoner species occur in good numbers at some time of the year. Post-breeding flocks begin to build up in late summer, when up to 40 Mute Swans and around 150 Tufted Duck and Great Crested Grebe are present. August and September are good months for Pintail and Garganey. Indeed, Blithfield is one of the most consistent sites for both species, with Pintail frequently reaching double figures. By October up to 50 Shoveler are present and other species are beginning to arrive. From then until February over 2,000 wildfowl are usually present, with almost 3,000 in November and December, when many species are at their peak. At this time there may be 500 to 600 Canada Geese, 600 Teal and 1,500 to 2,000 Mallard. A very few Gadwall are also regular throughout the winter. The most important dabbling duck is Wigeon. January and February counts frequently exceed 1,000 and even top 1,600 making this by far the most favoured regional site for this Russian winter visitor.

Wintering diving duck are more puzzling. Pochard flocks barely total 100, except around the year's end when they often reach 300 for a short while. Tufted Duck are also scarce, seldom numbering many more than 100. These meagre counts contrast sharply with those for Goldeneye, Goosander, Ruddy Duck and other diving birds. By January, 40 or so Goldeneye have normally assembled and most then remain until March or April. Sometimes an influx of birds on their way north in March leads to twice this number, by which time the resplendent drakes are busily displaying to the drabber ducks. Goosander also reach their peak after the New Year, when gatherings up to 80 make this the most important regional locality for this species too. Equally outstanding are the rafts of 400 Ruddy Duck in mid-winter. Indeed, in 1981 there were more than 600, which at that time was almost half the British population. Mid-winter also sees 500 to 600 Coot and some 80 Great Crested Grebes and Cormorants, though the latter quickly disperse after their early morning fishing expeditions.

Of the rarer wildfowl, Bewick's Swans and a few sea-duck pass through every year and may stay for some time. Indeed, a few Bewick's Swans, perhaps as many as 30 on occasions, are frequently present from late autumn until March. Though much rarer, Whooper Swans also visit. Of the sea-duck, Common Scoter is most regular, with small parties pausing briefly in spring, late summer and autumn during their overland migrations. A few Scaup are seen most years too, with some unusual summer records of late.

Other sea-duck, divers and the rarer grebes are more erratic visitors, whose arrival often coincides with late-autumn storms in the North Sea or extreme cold in the Baltic. Recently Black-throated and Great Northern Divers, Red-necked Grebe, Eider and Red-breasted Merganser have arrived during such conditions. Smew is irregular here, but one is seen from time-to-time, often in January. Small parties of grey geese also visit. Usually these are suspected of being feral birds, but now and again the gaggle size or general behaviour suggests wild birds. Parties of White-fronted Geese, particularly in December, may well be wild, while the occasional skein of overflying Pink-footed Geese in mid-winter almost certainly comes from the huge concentrations on the Lancashire mosses. Most intriguing of all, parties of around a dozen Barnacle Geese have twice appeared in recent Novembers.

Winter is also the time for gulls. Blithfield has a large gull roost that regularly holds one or two rarities. Black-headed Gulls predominate with 12,000 to 15,000, but both Lesser Black-backed and Herring Gulls regularly exceed 2,000 and Common and Great Black-backed Gulls 100 each. Numbers increase steadily during the autumn and by late November the first Glaucous Gull may appear. The scarcer Iceland Gull seldom arrives before December. The first arrivals of both species are usually immature birds and adults seldom appear until a month or two later. Numbers are variable. In a poor year only one or two Glaucous may be present, but a good year may see three or four Glaucous and two or three Iceland Gulls at a time. The higher numbers almost invariably occur during very cold weather, with February and March the most reliable months. Winter also brings mixed finch and bunting flocks, with maybe several Tree Sparrows and a few Brambling, while the alders are often visited by Siskin and Redpoll.

Spring passage begins in March, when winter visitors like Redwing and Fieldfare depart for northern climes. The end of the month often sees the first Sand Martins and Wheatears arrive, although most summer visitors do not show until April. Then many Pied and a few White Wagtails pass through, followed in May by parties of up to 80 Yellow Wagtails. With the latter might be a Blue-headed Wagtail. The steady stream of summer visitors periodically brings a scarcer migrant like Pied Flycatcher, or an exotic rarity like Spoonbill, Hoopoe or Golden Oriole. There is also a good chance of a passing raptor like Osprey, Hobby or even a Hen Harrier or Short-eared Owl.

Waders in spring are few in number, but very varied. Commonest are Ringed Plover, Dunlin and Common Sandpiper, though even these rarely reach double figures. However, two or three Oystercatchers, Sanderlings, Greenshanks and Turnstones regularly appear and some years a pair of Little Ringed Plovers attempt to breed. A Grey Plover, Black-tailed or Bar-tailed Godwit, Whimbrel, Green Sandpiper or rare Temminck's Stint could even add a touch of quality. Although the first waders may come in March, late April and early May is when most pass through. The same is true of gulls and terns. Kittiwakes are sometimes seen in March, while Little Gulls and terns regularly move through in late April and May. Arctic is the first tern to arrive, followed by Common. Neither is especially numerous, but if migrating Arctic Terns get caught up in a vigorous depression then well over 100 can occur along with a few waders and maybe an Arctic or Great Skua. Black Terns also occur, but are rather erratic, while most years bring a solitary Sandwich Tern in April or Little Tern in June.

As a breeding site, Blithfield is less outstanding. However, the colony of Great Crested Grebes is important, producing up to 40 young if nests are not flooded, while Little Grebe, Mallard and Tufted Duck are all increasing their productivity. Kingfishers nest nearby and in the adjoining woods Sparrowhawk, Tawny Owl, all three woodpeckers and a good range of woodland songsters breed, including perhaps a pair of Redstarts.

From July onwards waders become the prime interest. Although numbers are insignificant nationally, they are excellent for an inland location and outstanding for the West Midlands. Almost 40 species have been recorded, including national rarities such as White-rumped Sandpiper, Buff-breasted Sandpiper and Lesser Yellowlegs. July normally brings up to six Green Sandpipers, 20 Common Sandpipers and the first Snipe, but it is usually August before conditions become ideal. Then some 30 Ringed Plover and Dunlin may be expected, with very exceptionally Little Ringed Plovers just as plentiful.

Up to half-a-dozen Oystercatchers, Knot, Black-tailed Godwits, Whimbrel and Spotted Redshank usually pass through about this time too, along with slightly more Greenshank. Blithfield is very good for Ruff, with at least a dozen present at some time between July and September.

It is also the most reliable place in the region for Little Stint and Curlew Sandpiper. Most years bring just two or three of each, but occasionally a dozen or more appear. Among the less regular visitors might be a Sanderling, Wood Sandpiper or Turnstone.

After August there is a good chance of Grey Plover, a few Bar-tailed Godwits or even a Pectoral Sandpiper. By September some 150 Snipe have congregated, to be joined in October perhaps by a couple of Jack Snipe, at which time a Grey Phalarope may be swimming in the shallows. A small flock of Curlew and several thousand Lapwing may also have gathered. Dunlin too can be numerous in late autumn, with 50 or more in a tight, wheeling flock. If the weather stays mild a few might remain through the winter, accompanied perhaps by a Redshank or Common Sandpiper. Even a Purple Sandpiper has been known in mid-winter, though this is unusual.

*Blithfield is outstanding for passage waders such as Curlew Sandpiper (foreground), Dunlin and Ruff (background)*

August and September also see Little Gulls and terns return in small numbers, with again the chance of a party of Black Terns or a single Sandwich or Little Tern. Shortly afterwards the first autumn gales are awaited with high hopes. Seldom do they disappoint. A few Kittiwakes are almost guaranteed and a seabird is quite likely. Shags are commonest, but an Arctic Skua is quite probable while among the least expected in recent years have been Fulmar, Leach's Petrel, Sabine's Gull and Little Auk. Not all rarities are storm driven, however, and other autumn sightings have included Mediterranean Gull and Twite. By October you can count on a few Rock Pipits around the water's edge as well, while from November onwards a Snow Bunting is possible.

With so many waders and wildfowl, at least one Peregrine is now seen every autumn, usually in August, with frequent sightings thereafter until mid-autumn and then sporadic ones through the winter. August usually brings migrant Hobby and Osprey too, with sometimes a Marsh Harrier. Later, in October, a Hen Harrier might pass through.

## Timing

There is always something to see at Blithfield. For maximum numbers, especially of Cormorants, a visit soon after dawn is recommended, as they and some other species tend to move away as human disturbance increases during the day. The causeway gets very busy on sunny afternoons, but access to the rest of the reservoir is restricted, so disturbance should be minimal. Trespass is increasing, however, so it is best to visit as early as possible. Sailing is confined to south of the causeway and October to March is the close season for fishing.

## Access

The causeway, which crosses the reservoir, is here part of the B5013 Rugeley to Uttoxeter road. There is limited parking at either end of the causeway which is ideal for a quick visit, as many wildfowl and gulls can be seen from here. Parts of the reservoir can also be viewed from a minor road that skirts the eastern shore for a short distance. Serious birdwatchers are unlikely to find these views adequate, however, and to have a chance of seeing any waders or most of the better species it is advisable to obtain a permit from the West Midland Bird Club (WMBC). This gives access right round the reservoir and into the hides erected at strategic points by the South Staffordshire Waterworks Company. Without a permit there is strictly no access to the confines of the reservoir.

## Calendar

*Resident:* Little Grebe, Great Crested Grebe, Cormorant, Grey Heron, Mute Swan, Canada Goose, Mallard, Tufted Duck, Sparrowhawk, Kestrel, Coot, Lapwing, Little Owl, Tawny Owl, Kingfisher, all three woodpeckers and common woodland passerines.

*April–June:* Shelduck, Garganey (rare), Scaup (rare), Common Scoter (scarce), passing raptors such as Hen Harrier, Hobby, Osprey or Short-eared Owl (rare), Oystercatcher, Little Ringed Plover, Ringed Plover, Grey Plover (scarce), Sanderling, Temminck's Stint (very rare), Dunlin, Black- and Bar-tailed Godwits (scarce), Whimbrel (scarce), Curlew, Redshank, Greenshank, Green Sandpiper, Common Sandpiper, Turnstone, Little Gull, terns including a few Sandwich, Little and Black, hirundines, wagtails including White and occasional Blue-headed, Redstart, Wheatear, and common woodland warblers. Possible rarity.

*July–September:* Shelduck, Wigeon, Gadwall, Teal, Pintail, Garganey, Shoveler, Pochard, Common Scoter (scarce), Ruddy Duck, passage raptors such as Marsh Harrier and Osprey (rare), Hobby, Peregrine, Oystercatcher, Little Ringed Plover, Ringed Plover, Grey Plover (scarce), Knot, Sanderling (scarce), Little Stint, Curlew Sandpiper, Dunlin, Ruff, Snipe, Black-tailed Godwit, Whimbrel, Spotted Redshank, Redshank, Greenshank, Green Sandpiper, Wood Sandpiper (scarce), Common Sandpiper, Turnstone (scarce), Little Gull, terns including a few Sandwich, Little and Black, hirundines, Wheatears and common passage migrants.

*October–March:* Bewick's Swan, Whooper Swan (rare), grey geese (rare), Wigeon, Gadwall, Teal, Shoveler, Pochard, Scaup, Common

Scoter (scarce), Goldeneye, Smew (rare), Goosander, Ruddy Duck, Hen Harrier (rare), Golden Plover, Grey Plover (autumn), Dunlin (autumn), Jack Snipe (scarce), Snipe, Bar-tailed Godwit (autumn), Curlew (autumn), perhaps wintering Redshank, Green Sandpiper or Common Sandpiper, Grey Phalarope (rare in autumn), gulls including one or two Glaucous or Iceland in January–March, Kittiwake (autumn and March), Rock Pipit (October), Grey Wagtail, winter thrushes, mixed finch and bunting flocks including Tree Sparrows and a few Brambling, Siskin, Redpoll and perhaps Snow Bunting.

Possible vagrant seabird or rare gull in autumn; diver, rare grebe or sea-duck in winter.

## Habitat

Doxey Marshes reserve is owned by the Staffordshire Nature Conservation Trust (SNCT). It covers some 360 acres (146 ha) of the Sow Valley and is sandwiched between the A5013 and the Euston to Crewe railway line on the north-western outskirts of Stafford. There is a transition from dry meadows and hawthorn hedges, through washlands and marshes drained by a system of ditches, to reed-beds and open water. A scrape has recently been created to improve the habitat for waders and wildfowl, and a new hide overlooks this and the Tillington Flash. The Trust has recently acquired further land and more facilities are planned.

The richness and purity of the botanical life in the meadows and pools supports an abundance of insect life. This in turn provides a well-stocked larder for birds. Furthermore, brine seepage from old salt workings results in slightly saline conditions, especially in Boundary Flash, which inhibits freezing in all but the coldest weather. In winter too, much of the area is prone to flooding.

## Species

Birds that typically breed or winter in marshes are the main interest. Of special importance are the dozen or so breeding pairs of Snipe, whose drumming fills the air on summer evenings. With over 30 pairs, this is also a regional stronghold for Sedge Warblers. A dozen pairs of Moorhen and Coot, and one or two pairs of Redshank, also breed, and both Little Ringed Plover and Kingfisher have done so. Among the passerines, Yellow Wagtails, a dozen pairs of Meadow Pipits, a few pairs of Reed Warbler and some 40 pairs of Reed Bunting nest. Breeding wildfowl are also of interest, with broods of Little and Great Crested Grebes, Mute Swan, Canada Geese and Tufted and Ruddy Ducks to be seen.

By late summer, as the breeding warblers leave, the low water level exposes mud and this begins to draw in waders. Small numbers of Green and Common Sandpipers pass through in July and August, occasionally accompanied by a Wood Sandpiper. Ringed Plover, Redshank, several Ruff and Greenshank, and maybe Little Stint or Spotted Redshank are also likely around this time, with one of them perhaps lingering into autumn. September sees several summer visitors using the marsh to rest or feed on their way south. Whinchats flit restlessly along hedgerows or sit upright on fence posts. Late in the evening up to 200 Yellow Wagtails and over 1,000 Swallows and martins come from all directions to roost in the reed-beds. Quite often a Hobby swoops out of the darkening sky to play havoc with the gathering hirundines. Soon these species will be gone, to be replaced in November by up to 300 roosting Pied Wagtails.

October sees the start of a major influx of Snipe from the Continent and, by November, 200 are regularly probing the mud for food. (Counting can be difficult, but as many as 600 have been present on occasions.) With them come a few smaller, shorter-billed Jack Snipe. They are generally more skulking, but with luck one might be seen feeding out in the open in front of the hide. A dozen or so are normal, but exceptionally as many as 30 have been present. Several Water Rails invariably arrive around this time as well, and may be glimpsed slipping between the reeds. Many Snipe, Jack Snipe and Water Rails use the marsh only as a staging post to feed and rest before passing onwards, but most will stay throughout the winter unless frozen out by hard weather. This is equally true of the 1,000 or so Lapwings that feed across the marsh.

*In autumn Doxey attracts large numbers of Snipe and a few Jack Snipe (foreground)*

Autumn also sees wildfowl beginning to arrive. Teal are frequently most numerous, with as many as 200 up-ending in the shallows or loafing around the shoreline. At other times less than 50 are present and Mallard, at around 100, is the commoner species. Mute Swan, Wigeon and Shoveler are often present, but only in very small numbers, whereas a large flock of Canada Geese makes periodic visits. Diving duck are fewer in number, with seldom more than 20 or so Tufted or Pochard, perhaps twice as many Coot and a couple of Ruddy Duck. However, the area now appears to be attracting several Goosander.

Parties of Redwing and Fieldfare are very evident in the meadows and hedgerows from autumn onwards. At this time a Short-eared Owl or Hen Harrier may stay for a few days and, with luck, might be seen silently gliding just above the reed-tops. Kestrels hover above the marsh in search of voles and sometimes, late on a misty winter's afternoon, a Barn Owl drifts effortlessly past. It is worth scanning the hedgerows and fences too, as a Great Grey Shrike has been seen before now.

In the meadows, wintering flocks of Linnet, Goldfinch and Chaffinch may be joined by a few Brambling. Often there is a flock of 100 Tree Sparrows as well, and small parties of Siskin and Redpoll fidget through the alders.

Flooding often occurs in mid-winter and this can force Water Rail, Snipe and Jack Snipe into the open to feed. Indeed, Jack Snipe counts are often highest in February, although more birds are almost certainly present, but concealed, at passage times. Flooding might encourage a small herd of Bewick's or Whooper Swans to join the resident Mute

Swans for a while. Following freezing conditions elsewhere, even a Bittern might be glimpsed among the reeds, or a Water Pipit seen around the shore.

By April, summer visitors return and one or two waders such as Oystercatcher, Green Sandpiper or even Wood Sandpiper pause briefly on their way north. Before long, the marsh is once again alive with singing Reed Buntings and Sedge Warblers.

## Timing

Not particularly critical, but early morning or just before dusk are usually less disturbed and birds are more active.

## Access

Approach from Doxey Road in Stafford and park in the 'pay-and-display' car park in Chell Road (opposite the rear entrance of Sainsbury's). Follow the waymarked posts from Doxey Road along the footpath that follows the south side of the River Sow. These will lead to the centre of the marsh and the hide.

## Calendar

*Resident:* Little Grebe, Great Crested Grebe, Mute Swan, Canada Goose, Mallard, Tufted Duck, Ruddy Duck, Kestrel, Moorhen, Coot, Lapwing, Snipe, Kingfisher, Meadow Pipit and Reed Bunting.

*April–June:* Redshank, Yellow Wagtail, Sedge Warbler and Reed Warbler return to breed. Passage waders may include Oystercatcher, Green Sandpiper and Wood Sandpiper.

*July–September:* Hobby (rare), Swallow, Yellow Wagtail, Whinchat. Waders may include Little Ringed Plover, Ringed Plover, Little Stint, Ruff, Spotted Redshank, Redshank, Greenshank, and Green, Wood and Common Sandpipers.

*October–March:* Bewick's Swan (rare), Whooper Swans, Wigeon, Teal, Shoveler, Pochard, Goosander, Hen Harrier (rare), Water Rail, Jack Snipe, Barn and Short-eared Owls, Pied Wagtail, Fieldfare, Redwing, Great Grey Shrike (rare) and flocks of Tree Sparrows and finches, including a few Brambling, Siskin and Redpoll.

# BELVIDE RESERVOIR &
# CHILLINGTON LOWER AVENUE Map 10 OS map 127

## Habitat

Belvide is a canal-feeder reservoir situated in pleasant, low-lying, mixed farming country some 7 miles (11 km) north-west of Wolverhampton. It is leased from the British Waterways Board (BWB) by the West Midland Bird Club (WMBC). Immediately south of the A5 trunk road, it covers just 182 acres (74 ha) and has a shoreline of less than 3 miles (5 km). Thus this reserve is small enough for easy watching and recording, and offers good, close-range views.

The extreme north-easterly corner and the eastern shore comprise a brick dam, but, this apart, the shoreline is natural and shelves gently. Water comes from three small streams, two on the western side and one on the south. Agricultural run-off into these enriches the nutrients in the reservoir and periodically leads to profuse growths of blanketweed and algal blooms. There is also a luxuriant growth of amphibious bistort which provides both food and shelter for birds.

Water level is critical. A consistently high level, though good for wildfowl, is poorer overall for birds than a fluctuating level which exposes mud in spring and autumn. For such a small reservoir, the shoreline is remarkably varied. There are patches of oozy mud, rushes and grasses on the western and northern shores, great water-grass in the southern bays, and a large stand of bulrush in the south-eastern corner. Behind the shoreline extensive areas of grass and rush provide rough grazing for cattle and cover for nesting birds. In the extreme south-west, the West Marsh provides an area of rank, dense great water-grass, willowherb, nettle and rush that is used as a late-summer and autumn roost. Small areas of woodland also border the site.

A little under 1½ miles (2.4 km) south-east of Belvide is the Lower Avenue at Chillington. Part of the grand approach to Chillington Hall, it comprises a wide avenue of woodland with a good variety of broad-leaved trees, including a few beech and hornbeam towards the western end, and a rich and varied shrub layer. Halfway along is a bridge over the Shropshire Union Canal, which at this point is in a deep, wooded cutting.

## Species

Belvide's prime interest is as a breeding, moulting and wintering site for wildfowl. Indeed, it is the tenth most important reservoir in Britain and, with up to 5,000 birds on just 182 acres (74 ha), carries a higher density of birds in winter than anywhere else in the region. Add to this a large winter gull roost, passage waders and terns, and the promise of a rarity, and it is clear there is plenty of interest at all times.

Breeding wildfowl regularly include up to 30 pairs of Mallard, 20 pairs of Tufted Duck and 80 pairs of Coot, along with small numbers of Great Crested and Little Grebes, Shoveler and Ruddy Duck. Shelduck, Gadwall, Teal, Garganey and Pochard have bred as well, though less regularly. A few pairs of Snipe breed in the adjoining pastures and the delightful courtship and territorial display flights of Redshank are a welcome sight in spring. Up to seven pairs may attempt to breed, but seldom more than a couple are successful.

One or two pairs of Curlew nest on nearby farmland and are regularly seen or heard. The West Marsh, plus perhaps the areas of bulrush, hold a few pairs of Reed Warbler and possibly even a secretive pair of Water Rails. Sedge Warblers are never numerous, but half-a-dozen pairs take up residence around the reservoir. The surrounding hedgerows and copses hold several pairs of Lesser Whitethroat, Blackcap, and Willow Warbler, plus small numbers of Whitethroat, Garden Warbler and Chiffchaff.

Despite its small size and disturbance from anglers in boats, Belvide still holds large flocks of moulting wildfowl. Up to 500 Mallard, mostly drakes, arrive around the beginning of June, when many ducks are still with their ducklings. Similar late-summer influxes of drake Pochard and Tufted Duck are noticeable, with up to 800 of the latter species, while the moulting flock of Coot may be 2,000 strong. Several thousand Swifts and hirundines descend on the reservoir to feed during the summer, especially seeking low-flying insects in poor weather. In late summer, several hundred Swallows and a few Sand Martins and Yellow Wagtails roost in the West Marsh.

Wader passage begins in late June, peaks in August and then slowly subsides and peters out in November. First to show are often Little Ringed Plover, which may reach double figures, Dunlin and Green and Common Sandpipers. August then sees the main passage of Ringed Plover and Ruff, both of which may well reach double figures. Two or three Greenshank are usually present at this time too, while any combination of Oystercatcher, Whimbrel, Black- and Bar-tailed Godwits, Spotted Redshank, Wood Sandpiper, Sandering, Knot and Turnstone might appear. Little Stint is seen most years, often juveniles in September, but Curlew Sandpiper is less regular. Finally, October or November may bring a solitary Grey Plover and, if conditions are right, a good flock of up to 80 Dunlin. Autumn also holds promise of a small passage of Arctic, Common and Black Terns, plus perhaps a Little or Sandwich Tern. Usually one or two Little Gulls come as well. Seasonal gales can enliven things considerably, bringing with them the chance of an Arctic Skua or even a rare and exquisite Sabine's Gull.

From late October onwards wildfowl steadily increase. The first dabbling duck to peak, in October and November, is Shoveler. At least 100 are likely and well over 500 have been recorded. There are then successive influxes of Mallard which culminate in a typical early- or mid-winter maximum of 1,300, but in a good year twice this number. Teal, too, peak around this time, followed by Wigeon a month or two later. Up to 300 Teal are likely, but Wigeon numbers have been poor recently, seldom exceeding 100. A few Gadwall are now regular, especially in spring, when small numbers of Shelduck are often seen as well.

Turning to diving duck, Belvide's speciality must be the Ruddy Duck. In addition to the small breeding population, a large flock about 300 strong gathers each autumn. Many quite often leave for Blithfield Reservoir at the end of the year, but a good number remain at Belvide if the winter is mild. Conversely, Goldeneye do not reach their peak of about 100 until the New Year, but most then remain through to mid-April. The winter peaks for Pochard and Tufted Duck are around 100 and 200 respectively, but up to 300 and 600 are present on rare occasions. Sawbills are currently scarce, although all three species have occurred in recent years. Belvide was once a recognised site for Goosander, but nowadays half-a-dozen is all that can be expected. Other wintering wildfowl include some 300 Coot, and a few Little and Great Crested Grebes, with the latter increasing to a spring peak of 100. Such large numbers of wildfowl inevitably bring one or two scarce species. Black-necked Grebe is now almost an annual spring and early-autumn visitor, as is Garganey in late summer, while a few Bewick's Swans, Barnacle Geese, Pintail, Scaup and Common Scoter are noted most years, and both Red-throated and Great Northern Divers have recently occurred.

*Belvide's speciality is the Ruddy Duck, shown here with Tufted Ducks beyond*

Winter also brings a reasonable gull roost, with up to 20,000 Black-headed, 1,000 Lesser Black-backed and 500 Herring Gulls. Amongst them may be a rarer gull, such as Mediterranean in autumn, or more likely Iceland and Glaucous in late winter. In addition to the wildfowl and gulls, a Bittern might appear in hard weather. Short-eared Owl and Firecrest have also occurred in winter, when Tree Sparrows, finches and buntings feed on open ground. With them may be a few Brambling or a Snow Bunting.

By March wildfowl are departing, though one or two Wigeon and Goldeneye often stay well into May or even later, and Curlew and Redshank herald the onset of spring passage. Small parties of Meadow Pipits frequently follow and the first Wheatears are soon flitting along the dam. Early April brings hirundines and wagtails. Prior to their population crash, Sand Martins were numerous, but now only meagre numbers arrive. Yellow Wagtails pass through in much better numbers, and with them usually come one or two White Wagtails and maybe a Blue-headed. There is usually some mud in spring, so wader passage is better than at many reservoirs. Generally it brings both Ringed and Little Ringed Plovers, Sanderling, Dunlin, Whimbrel and Common Sandpiper, and less often Black-tailed Godwit, Spotted Redshank, Greenshank or a sizeable party of Bar-tailed Godwits. Generally speaking, there is less variety in spring than autumn and visits are more fleeting, but the chances of a rarity are greater. Recently Kentish Plover, Red-necked Phalarope and Spotted Sandpiper have all appeared.

Terns also pass through in spring. Numbers are generally small, but deep depressions in late April or early May can bring up to 100 Arctic Terns. Common, and less predictably Black, Terns usually come later, and sometimes a Little or even a rare Roseate Tern appears. Spring is also good for raptors, with recent records of Marsh and Montagu's Harrier, Osprey, Hobby and Peregrine; and also a chance of a Spoonbill.

Hawfinches are the main attraction of the Lower Avenue, with small numbers present on and off throughout the year. Winter and more especially early spring are the best times to try. You can then sometimes see them feeding in the leaf litter beneath their favourite hornbeams and beeches. Once startled, however, they rocket into the canopy, where they are much harder to watch. Check the adjoining hedgerows too, and listen for their explosive calls. Lower Avenue is also good for resident woodland species like Great Spotted Woodpecker, Nuthatch, Tree-creeper and tits, for warblers in summer, and for flocks of Fieldfare, Redwing and finches in winter.

## Timing

There is little disturbance, except from anglers' boats. Water level and weather are more important. The former depends on BWB activities, but in spring is usually lower than most water-supply reservoirs, making Belvide a good place for waders. A visit after gales or storms can often prove productive, while waders and terns frequently appear during passing depressions. In very cold weather the reservoir virtually freezes over and few birds remain.

Lower Avenue can be busy at weekends and Bank Holidays, so for the shy, elusive Hawfinch an early morning or weekday visit is recommended.

## Access

As Belvide is leased by the WMBC access is strictly by permit only. (Permits are issued by the WMBC and are valid for one year.)

The reservoir is adjacent to the A5 Cannock to Telford road, but at a higher level, so it cannot be viewed from any surrounding roads. There is a track along the western boundary that is a public footpath, but on no account should cars be taken down this as it is a private road. In any case views from here are frustratingly restricted.

Permit holders should approach from the A5, turning southwards towards Brewood a little under 2 miles (3 km) west of the A449 roundabout junction at Gailey. Take the third turning left (a lane) off the A5 after this junction. Continue down this lane for ⅔ mile (1 km) and then turn right into another narrow lane that leads to Shutt Green. Cross over the canal, go through the tiny village and there is a car park on your right in a field adjoining a small wood. The gate is locked, but keys are issued to permit holders. Park here and walk through the wood to the reservoir and hides, which are also locked. To avoid disturbing ground-nesting birds visitors are asked to keep off the northern shore during the breeding season.

For Chillington Lower Avenue, return through Shutt Green to the end of the lane and turn right into Brewood. Follow signs out of the village for Coven. After ½ mile (0.8 km) turn right at the crossroads into Codsall Road. In ½ mile (0.8 km) turn left into another lane and park on the verge where the Lower Avenue footpath strikes off to the left. Remember to try the less used footpath to the right of this road as well.

# Calendar

*Resident:* Little Grebe, Great Crested Grebe, Mallard, Shoveler, Tufted Duck, Ruddy Duck, Coot, Snipe and common passerines, plus Great Spotted Woodpecker, Treecreeper, Nuthatch, Hawfinch and common woodland birds at Lower Avenue.

*March–June:* Black-necked Grebe (scarce), Shelduck, Garganey, Little Ringed Plover, Ringed Plover, Sanderling, Dunlin, Whimbrel, Curlew, Redshank and Common Sandpiper. Maybe Black- and Bar-tailed Godwits, Spotted Redshank or Greenshank. From mid-April: terns including possible Little and Black, Swift (from May), hirundines, Meadow Pipit, Yellow Wagtail, Pied and White Wagtails, Wheatear, warblers including Sedge and Reed, and other common summer visitors. Passing raptors. Maybe rarity.

*July–September:* Black-necked Grebe (scarce), Garganey, Pochard, common passage waders plus Ruff, Greenshank and possible Oyster-catcher, Knot, Little Stint, Curlew Sandpiper, Black- and Bar-tailed Godwits, Whimbrel, Spotted Redshank, Wood Sandpiper and Turnstone. Little Gull, terns including possible Sandwich, Little and Black, and departing summer visitors. Passing raptors. Maybe rarity.

*October–February:* Shelduck, Wigeon, Gadwall, Teal, Pochard, Goldeneye, Goosander, Water Rail, Grey Plover (October or November), gulls including possible Iceland and Glaucous (after December), and winter thrushes. Maybe a vagrant seabird after autumn gales, or possible diver or sea-duck in winter. Perhaps a rarity.

## Habitat

The 16,500 acres (6,700 ha) of Cannock Chase form the smallest Area of Outstanding Natural Beauty (AONB) in Britain. A magnificent expanse of semi-natural oak and birch woodland, upland heath and conifer plantations, the Chase is owned largely by the Forestry Commission (FC) and Staffordshire County Council (SCC). On fine summer weekends it attracts literally thousands of visitors. Indeed, pressure became so great that parts are now closed to cars in order to prevent erosion.

An upland plateau dissected by narrow valleys, the Chase reaches a height of 800 ft (243 m) at its northern end, from where beautiful, wooded slopes descend into the Trent Valley. Capped by Bunter sands and gravels, the soils are acidic and poor. Browsing by five species of deer, and (formerly) grazing by sheep, have inhibited natural regeneration and encouraged an open, heath flora, with extensive sweeps of bracken. Tiny streams meander along the valleys and mires have developed where drainage is impeded. Some valleys are quite distinctive, with the Oldacre Valley lined with willows and the Sherbrook Valley with old, coppiced alders.

Precious broad-leaved woodland still survives. There are pedunculate oakwoods at Seven Springs and Sycamore Hill, and a superb sessile oakwood at Brocton Coppice. The Brocton oaks are mostly 200 to 300 years old, but their invertebrate fauna suggest a woodland with much more ancient origins. With reduced grazing, birch, sycamore and Scots pine are freely regenerating and hawthorn, crab apple, holly, rowan, willow and elder are all becoming steadily established. Plantations of pine, larch and spruce now cover much of the southern and eastern sides of the Chase. With maturity, they are now more attractive to birds.

Standing water is scarce, but there are a few small pools within the plantations where birds sometimes drink and bathe. Waterfowl are few, however, except in the flooded Brocton Quarry with its towering exposure of Bunter pebble beds.

## Species

Scrub, woodland and even wetland birds are all interesting, but most birdwatchers are seeking heathland species, particularly Nightjar.

Meadow Pipit and Linnet breed commonly on the heath and up to 50 pairs of Tree Pipit are spread widely across areas with scattered birches and oaks. Here, too, Willow Warblers, Chaffinches and Yellowhammers are abundant and the laughing calls of Green Woodpeckers echo across the heather. When crossing the heath watch for Whinchats perched on bracken fronds or a scolding Stonechat on a gorse bush. Both are regular passage migrants and a few pairs breed most years.

The full glory of the heath is best savoured on a balmy, summer's evening. As the sun sinks, so Grasshopper Warblers start their incessant reeling from deep within a tangle of vegetation. Soon Woodcock add their 'squeaks and groans' during their roding flights.

They can be seen in many, widely scattered, heathland localities. Finally, when only silhouettes can be seen against a darkened sky, Nightjars begin to churr. With some 25 pairs the Chase is unquestionably the Midland's stronghold for this species.

*At dusk on the Chase, Nightjars hawk insects and Woodcock rode above the trees*

In late summer passage Wheatears and Whinchats visit the heath, followed in early autumn by occasional Stonechats and Ring Ouzels. The latter favour the rich harvest of rowan and elder berries in the Oldacre Valley. As the first autumn frosts begin to bite, flocks of Fieldfare and Redwing descend on the same food source. With luck a passage raptor may pause for a few days, or even stay for the winter. A ring-tailed Hen Harrier is most likely, but Merlin and Short-eared Owl are also possible. A more reliable winter visitor is Great Grey Shrike, one or two of which come each year. Indeed, the Chase is really the only regular regional haunt. As they wander quite widely, location can be difficult, but a favoured spot is by the glacial boulder on Chase Road. Sooner or later a sighting in this vicinity is likely. This is also a good spot for a harrier.

The birdlife of the broad-leaved woodlands is extremely good. Among the commoner species, Tawny Owl, Great Spotted Woodpecker, Nuthatch, Treecreeper, Jay and a good range of tits and finches are resident, while summer visitors include Chiffchaff, Willow Warbler, Blackcap, Garden Warbler and Spotted Flycatcher. A few pairs of Wood Warbler nest where the shrub layer is sparse, with Brocton Coppice and Seven Springs likely areas. Redstarts and Pied Flycatchers breed sparingly too, with the old oaks of Brocton Coppice again a favoured haunt. Small numbers of Willow Tits and Lesser Spotted Woodpeckers occur among the valley willows and alders, especially along the Sherbrook Valley. In winter these same alders are visited by twittering flocks of Goldfinch, Redpoll and Siskin, and parties of foraging tits, Goldcrests and a few Treecreepers.

The Forestry Commission (FC) has planted some 6,700 acres (2,700 ha) with conifers. These are old plantations, with serried ranks of larch, spruce, Corsican and Scots pines surrounded by a fringe of birch, beech and oak. Now, some 60 years after planting, the oldest are being clear-felled and replanted, creating a wider age structure to the benefit of birds. Goldcrests and Coal Tits are both widespread and abundant, their high-pitched calls a characteristic sound throughout the year. Always check parties of Goldcrests carefully as Firecrests have been seen both in spring and autumn. Woodpigeons are another familiar sight flying to and fro above the forest canopy. Redpolls, too, are numerous and can usually be located by their metallic trills. The heart of the plantations are too dark and lifeless to harbour many birds. However, Great Spotted Woodpeckers feed on old decaying tree stumps, Chaffinches, Blue and Great Tits, Robins and Dunnocks frequent the woodland fringe, and Reed Buntings nest alongside Yellowhammers in the bracken-covered rides.

From July onwards the 'chip, chip' calls of Crossbills may be heard as irrupting parties fly overhead or feed in the dense canopy. Although larch, pine and spruce cones are their main food, birds on the Chase have frequently used young oaks as perches and song posts. Crossbills are recorded virtually every year and have bred on occasions. Once, a brightly-coloured male Two-barred Crossbill delighted birdwatchers for almost four months, so always closely scrutinise any Crossbills present.

In springtime large flocks of Siskin, sometimes up to 400 strong, feed among spruce and pines. Most soon leave to spend the summer further north, but increasingly a few linger later each year and have even bred. A third finch to look for is Brambling. Seldom numerous in the Midlands, small numbers sometimes roost with Chaffinches and occasionally a late male may sing in April or May.

Birds of prey are well represented too. Sparrowhawks are the commonest and on bright days several might spiral together on a rising thermal. From time-to-time a larger raptor, such as a Buzzard, passes over, usually being mobbed by the local crows or Sparrowhawks. Within the plantations, some half-a-dozen pairs of Long-eared Owl nest, but the chances of finding this essentially nocturnal bird are slim. On a still, summer night, however, the 'squeaky-gate' calls of hungry youngsters might just be heard.

Moorhen and Mallard inhabit the small pools, but most waterfowl in the area are to be found on Brocton Pool. Great Crested Grebe, Little Grebe, Tufted Duck and Kingfisher all breed here and a small flock of Canada Geese is regularly present.

## Timing

On fine summer days the Chase is thronged with visitors, though few venture far from the car parks. Nevertheless, afternoons are best avoided. For the woodland birds, try first thing in the morning when they are most active. An evening walk across the heath is also a good idea, as Grasshopper Warblers sing better then, Woodcock begin to rode at dusk and Nightjars are seldom seen or heard until it is virtually dark. They are most active on warm evenings when there are plenty of flying insects. In winter, weather is most important. Avoid wet or windy conditions. Cold, clear mornings with a light sprinkling of snow can be good, as birds are actively feeding to regain the body-weight lost overnight. The stands of conifers are fairly quiet and undisturbed at most times.

## Access

Cannock Chase lies between the A34 Cannock to Stafford and the A513 Stafford to Rugeley roads, and either side of the A460 Cannock to Rugeley road. The heathland has a well-defined pattern of peripheral car parks and footpaths into the area. The FC plantations can be explored on foot using the many forest roads. The best access points are:

*Brocton Field and Coppice* From the north leave the A34 eastwards 2½ miles (4 km) south of Weeping Cross, Stafford, where it is signposted 'German Cemetery and Hednesford'. In a further 1 mile (1.6 km) turn left onto a metalled road with speed ramps. This leads across the Chase and there are plenty of parking places from where paths radiate across the heath. For Brocton Coppice and Sherbrook Valley turn right by the old gravel pit. The coppice is then straight ahead and easy paths lead off right into the valley.

*Oldacre Valley and Brocton Pool* Turn southwards in Brocton village, then right into Oldacre Lane, where there is limited parking at the end of the road. Continue on foot across the old gravel workings to Brocton Pool, where a public hide overlooks the water. A track round the east of the pool goes out into Oldacre Valley.

*German Cemetery and Sherbrook Valley* Leave the A34 2½ miles (4 km) north of Cannock by turning eastwards at the crossroads towards Rugeley. In 1 mile (1.6 km) turn left at the first crossroads and a short distance along turn right into the car park adjoining the German Cemetery. From here follow the main footpath into Sherbrook Valley.

*Seven Springs* Approach from the A513 Stafford to Rugeley road, taking a narrow track westwards opposite the turning to Little Haywood. Follow this track to the car park among birch trees and then explore on foot.

*Beaudesert Old Park* This is one of the better areas of coniferous woodland. Leave the A460 Cannock to Rugeley road to the east 1 mile (1.6 km) south of Rugeley. Continue for 1¼ miles (2 km) to Wandon crossroads, park on the verge and follow the paths and rides on foot. The area to the south and south-east is often good for Crossbills and Siskin.

## Calendar

*Resident:* Little Grebe, Great Crested Grebe, Canada Goose, Mallard, Tufted Duck, Sparrowhawk, Woodcock, Tawny Owl, Long-eared Owl, Kingfisher, all three woodpeckers, Skylark, Meadow Pipit, and common heath and woodland birds including Goldcrest, Willow Tit, Nuthatch, Treecreeper and Redpoll.

*April–June:* Nightjar (after mid-May), Tree Pipit, Redstart, Whinchat, Stonechat, Grasshopper Warbler, woodland warblers including Wood, Spotted and Pied Flycatchers, Siskin and Brambling in April. Perhaps a migrant raptor.

*July–September:* Stonechat, Whinchat, Wheatear, Ring Ouzel, and other departing summer visitors. Maybe an irruption of Crossbills.

*October–March:* Fieldfare, Redwing, Great Grey Shrike, and flocks of tits and finches including Siskin, Redpoll and a few Brambling. Possible Hen Harrier, Merlin or Short-eared Owl.

## Habitat

Chasewater is a British Waterways Board (BWB) canal-feeder reservoir of almost 250 acres (100 ha) set in a landscape of former heath, marsh and bog. Over the years this landscape has been steadily eroded by mining and encroached upon by the factories and houses of Brownhills, Chasetown and Norton East. Recently, an interesting mosaic of marsh and scrub was reclaimed as a monoculture grassland. Now, following initial leisure development begun 30 years ago, it is proposed to develop a major leisure park complex. Fortunately the plans for this are very sympathetic towards nature conservation, but inevitably some good habitats are threatened and the nature of the site would change. The proposal is currently being jeopardised by financial constraints and the threat of a new motorway along the southern shore of the reservoir, so the future remains uncertain. Meanwhile local youths persist in using the area for motor-cycling and air rifle practice.

The central feature of the Chasewater site is the reservoir itself, which is divided into two by a causeway. This once carried a railway line into several, now long-defunct, collieries. To the south of this causeway, the main body of water, an area of 217 acres (88 ha), is used for fishing, sailing, windsurfing, water-skiing and power-boat racing. This disturbance makes the much smaller body of water to the north, known as Jeffrey's Swag, especially important as a quiet refuge for wildfowl.

From the south, the approach to Chasewater is via the A5. You are confronted by a stadium and trotting track on your right and an amusement park along the southern shore to your left. Do not be deterred. Despite the deafening go-karts and general hurly-burly, this is often the shoreline that waders, pipits and wagtails prefer! Around the western shore, the reclaimed grassland is good for passage Wheatears, but little else, while below the embankment the small pools, scrub and fields of Willow Vale are also of interest.

On the opposite (eastern) side of the reservoir is the dam. Beneath this, Anglesey Basin and the canal-feeder are bordered to the south by dry heath and to the north by a richly varied marsh noted for its passage migrants and winter roosts. The heathland continues north of the dam, where it is interspersed with scrub, marsh, woodland, a small pool and open ground. Together these make an excellent habitat for birds, with a good variety of passerines, unlike the reclaimed grasslands beyond, where only the alder plantations are of interest. Despite constant disturbance and damage by motor-cyclists, the shoreline between the dam and Target Point is good for waders, but beyond the Point it becomes peaty and less attractive.

North of the disused railway line are old pit mounds and slurry beds, where motor-cyclists scramble. Despite this, it still holds interesting birds and ought not to be missed. It was a favoured area of wintering Twite until disturbed by re-excavation of the slurry beds. Now there is a pool and marsh suitable for breeding duck and waders, while the dense willow scrub around Plant Swag is often used by roosting wagtails. Similar dense patches of willow along Big Crane Brook and around. Jeffrey's Swag shelter breeding and passage warblers, while the patches of gorse and broom on the adjacent heath are the most likely places for chats.

The small stream flowing into Jeffrey's Swag from the north used to regularly flood, and was a well-known haunt of Water Rail and Jack Snipe. Since being dredged, however, it no longer floods and so has lost much of its interest. Further north, on Southacres Farm, large areas of wormwood attract flocks of wintering finches, including perhaps a few Twite.

## Species

Winter wildfowl and gulls, and passage waders and terns are the main interest, but above all Chasewater is known for its rarities. Wildfowl numbers increase from August onwards and for most species reach their peak in December. After this there is a steady decline, until by April few birds are left. Diving duck and Coot predominate, with typical maxima of 500 for Tufted Duck and Coot, and 80 to 100 for Goldeneye. Unlike other species, Goldeneye peak later in the winter, usually in March. The water is much less attractive to Pochard, and although 50 or so may arrive with the October influx into Britain, fewer than this remain through the winter. Of the other diving duck, one or two Goosander and Ruddy Duck can also be expected, while Scaup and Common Scoter occur most years, the latter sometimes in small flocks in spring, late summer or autumn. Scarcer species such as Eider, Velvet Scoter, Long-tailed Duck, Smew and Red-breasted Merganser are unpredictable, but all have occurred several times. The most likely times for them to arrive are in very hard weather or in the aftermath of severe gales.

One or two Great Crested Grebes are present throughout the year, with most in late spring and autumn. Little Grebes are usually fewer and tend to frequent the smaller pools in preference to the main reservoir, particularly during the breeding season. At some time during the winter a rarer grebe or diver is likely. In recent years, Black-necked, Red-necked and Slavonian Grebe and all three divers have been seen, with this site an especially favoured haunt of Black-throated Diver.

Dabbling duck are less numerous, with typically only 200 or so Mallard, 10 Teal and Wigeon, and a few Gadwall and Shoveler. Migrating Shelduck pass through, mostly in April, September and December. Pintail are scarce and rather unpredictable. There is a small herd of resident Mute Swans, which increases to a peak of 20 to 30 between September and January, while between October and March a small party of Bewick's Swans may pause briefly to rest and feed. Whooper Swans, too, occur from time-to-time. They are very scarce, but increasing.

As winter progresses and wildfowl numbers begin to decline during February and March, attention turns to gulls. Having spent the day gorging on farmland or rubbish tips, thousands flock into the reservoir every evening to bathe and roost. They are there from autumn through to spring, though the rarer gulls are more likely in late winter.

Sometimes power-boats and water-skiers disturb roosts, causing many birds to leave, most probably for Blithfield Reservoir. When undisturbed, however, this is the best place in the region to study gulls. Black-headed Gulls are by far the most numerous, usually peaking at around 12,000 in December and January. The proportion of larger gulls is above average, though, with 2,000 Lesser Black-backed and Herring Gulls and 100 Great Black-backed likely. Amongst these, two or three Glaucous and one or two Iceland Gulls are regularly present. Indeed,

Chasewater has to be one of the best and most consistent inland sites in Britain for these two species. Up to 50 Common Gulls are also present and the rare Mediterranean Gull has been seen.

*Adult Iceland Gull (left) and immature Glaucous Gull are typical of the unusual gulls for which Chasewater is renowned*

Spring wader passage may begin with one or two Oystercatchers, Little Ringed Plovers, Ringed Plovers or Dunlin in March, but most movement is concentrated into a few frenetic days in May, when small parties of Ringed Plover, Dunlin and Common Sandpiper, one or two Sanderling or Turnstone, and perhaps a party of Whimbrel pass swiftly through. Sanderling in particular favour the shoreline near the sailing club.

Return passage is more protracted and sedate. The water level is kept as high as possible for water sports, but if it does fall the exposed muddy margins attract a good variety of waders, mostly along the southern shore or east of Target Point. Again the commoner species are the more likely, particularly Ringed Plover, Dunlin and Common Sandpiper. These are usually joined in August by Ruff and Greenshank, and in September by one or two Curlew and Knot. Less often, Grey Plover, Little Stint, Curlew Sandpiper or Bar-tailed Godwit appear, while exceptionally a rarity enlivens the day.

Unlike waders, terns and gulls are unaffected by water level. Spring passage sometimes begins with a few Kittiwakes in March, but, as with waders, it reaches a crescendo in a few hectic days, or even hours, in late April or early May. Most birds pass quickly through, with terns in particular seeming to appear and disappear 'out-of-the-blue'. During deep depressions numbers can be impressive; over 500 Arctic Terns and a flock of 27 Little Gulls have been recorded – both regional records. Common Terns are regular, but less numerous than Arctic, Black Terns are erratic and both Little and Sandwich Terns are increasing. Autumn usually brings more records, but fewer birds, than spring.

The surrounds of Chasewater are almost as interesting as the reservoir itself. Woodpeckers can be seen or heard throughout the year, while in summer the heath supports many Skylarks, Meadow Pipits and Willow

Warblers. With luck a Grasshopper Warbler might even be heard. Jays and Redpolls frequent the scrub, and passage Wheatears and the occasional Whinchat flit across the heather or perch upright on convenient gorse bushes. In winter the damp hollows and marshy areas hold good numbers of Snipe, a few Jack Snipe and one or two Water Rails, while a Short-eared Owl might be seen quartering the heath in search of voles. Indeed, the heath is the best spot for raptors, with Sparrowhawks regularly overhead. Meadow Pipits and passage Wheatears are two of the few species that frequent the newly reclaimed grasslands, but the alder plantations are visited by wintering flocks of Siskin and Redpoll.

Do not overlook the marsh behind the dam. Here, amidst dense willow scrub and rank vegetation, Whitethroat, Sedge, Reed and Willow Warblers, Willow Tit, Bullfinch and Reed Bunting are among the regular breeding species, while Water Rail and Whinchat nest irregularly. Later there is a steady stream of migrant warblers, especially Willow Warblers in late summer, followed in autumn by flocks of Linnets and Goldfinches, and in winter by Snipe and Water Rail. Several species also use the willows as a winter roost. Across the reservoir, Willow Vale might also hold Water Rail and visiting Hobbies often approach from this direction.

Along the old railway line the tall thorn scrub holds breeding Lesser Whitethroats and is used by Jays moving to or from the nearby Brownhills Common. In winter the surrounding pastures are visited by flocks of Fieldfare and Redwing, or Skylarks, Tree Sparrows and finches. These often include Redpoll and sometimes a few Brambling, particularly in early spring. A real speciality of Chasewater used to be its flock of 100 or more wintering Twite. Sightings and numbers have regretfully dwindled since re-excavation of the slurry pits disturbed a favoured feeding ground, but a few are still seen. Try the old pit mounds, where there are also Grey and Red-legged Partridge, or watch from the edge of Southacres Farm for finches going in and out of the wormwood. Remember, though, there is no access to this private farmland.

Winter may bring other interesting species too. Sometimes a Merlin chases small birds across the heath, or a Long-eared Owl roosts in a willow thicket. Another time a Water Pipit will feed around the shoreline or a few Snow Buntings will arrive like snow flakes in November. Of the real rarities, Cory's Shearwater sitting exhausted on Target Point, Lesser Kestrel, Red-footed Falcon on the heath, Least Sandpiper, Buff-breasted Sandpiper feeding by the sailing club, Caspian Tern and White-winged Black Tern are outstanding for any site, leave alone one so far inland.

## Timing

There are always birds worth seeing, except perhaps in high summer when disturbance is at its worst. Then, by way of compensation, there are many orchids, sundews, dragonflies and butterflies to be enjoyed. Early in the day is the best time to visit, as water sports and motor-cycle activity both intensify in the afternoons. If there is disturbance on the main reservoir in winter, remember to check Jeffrey's Swag for wildfowl. Duck numbers tend to be higher in cold weather, when smaller, shallower waters are frozen over. In spring, passerines tend to arrive with southerly winds and overcast conditions, but storms or deep depressions often bring waders and Arctic Terns. Black Terns and Little Gulls, on the other hand, tend to arrive on light easterly winds in settled weather. Vagrant seabirds are most likely after autumn gales, while divers or the rarer grebes frequently come with very cold weather. The colder nights also seem to be the better ones for Glaucous and Iceland Gulls.

## Access

The main approach is from the A5, which at this point is a dual carriageway. There is access only from the eastbound carriageway, so when travelling westwards proceed past the entrance and on to the roundabout at the junction with the A452. Go right around the roundabout and retrace your route, but on the eastbound carriageway. From this carriageway turn left into Pool Road, just before the traffic lights, towards the reservoir.

In winter, turn left again into the amusement park. The car park here overlooks the southern shore, so the gull roost can be studied from the comfort of your car. In summer a charge is made for entry to the amusement park, but this can be avoided by carrying straight on past the entrance to the end of the dam, where there is limited parking on the verge. There are well defined and obvious footpaths around the reservoir and its surrounds.

Jeffrey's Swag and the northern area can be reached from Norton East, where cars can be parked along the road.

## Calendar

*Resident:* Little Grebe, Great Crested Grebe, Mute Swan, Mallard, Tufted Duck (rare in summer), Sparrowhawk, Kestrel, Red-legged Partridge, Grey Partridge, Water Rail, Coot, Snipe, Green and Great Spotted Woodpeckers, Skylark, Meadow Pipit and the usual common passerines, including Willow Tit, Jay, Redpoll, Bullfinch and Reed Bunting.

*April–June:* Shelduck, Hobby (scarce) or other passage raptors, Oystercatcher, Ringed Plover, Sanderling, Dunlin, Whimbrel, Common Sandpiper, Turnstone, Little Gull, terns including occasional Sandwich, Little and Black, hirundines, Whinchat, Wheatear, Grasshopper Warbler, Sedge Warbler, Reed Warbler, Lesser Whitethroat, Whitethroat, Blackcap and Willow Warbler. Possible rarities.

*July–September:* Hobby (scarce) or other passage raptors, Little Ringed Plover, Ringed Plover, Grey Plover (scarce), Knot, Little Stint (scarce), Curlew Sandpiper (rare), Dunlin, Ruff, Bar-tailed Godwit (scarce), Curlew, Redshank, Greenshank, Common Sandpiper, Turnstone, Little Gull, terns including occasional Sandwich, Little and Black, hirundines, Whinchat, Wheatear and departing summer warblers. Possible rarities.

*October–March:* Cormorant, Bewick's Swan (scarce), Shelduck, Wigeon, Gadwall (scarce), Teal, Pintail (scarce), Pochard, Scaup (scarce), Common Scoter (scarce), Goldeneye, Goosander, Ruddy Duck, Merlin (scarce), Jack Snipe, gulls regularly including Iceland and Glaucous (after mid-November), Kittiwake (autumn and March), Long- and Short-eared Owls (scarce), Rock Pipit (October), Water Pipit (rare), Grey Wagtail, Fieldfare, Redwing, winter flocks of passerines including Tree Sparrow, Brambling, Siskin and Twite (very scarce). Possible Snow Bunting (November). Maybe vagrant seabird or storm-driven diver, rare grebe or sea-duck. Possible rarity.

The Black Country lies roughly between Wolverhampton, Walsall, West Bromwich and Dudley. During the Industrial Revolution countless settlements sprang up around a multitude of mines, mills, factories and quarries. Before long these were firmly established as the 'workshop of the world'.

Villages grew into towns, and the towns spread outwards until they met. Somehow a few pockets of undeveloped land survived, increasingly isolated from the countryside beyond. Some of these, prized and protected as urban lungs, now provide some fascinating habitats. Two examples are Sandwell Valley and Saltwells Wood, the latter, ironically, adjacent to one of the first government Enterprise Zones.

Heavily dependent on traditional manufacturing industry, the Black Country has suffered badly from the industrial recession of the 1980s. Today there are countless acres of derelict land, much of it too toxic for wildlife. The more interesting sites are detailed below.

## SANDWELL VALLEY

### Habitat

Sandwiched between Birmingham to the east and West Bromwich to the west is a large lung of open countryside that might be many miles from either. For here the pleasant countryside is dotted with woods and lakes, but unfortunately in the very centre is the M5/M6 motorway interchange. The heavily polluted River Tame also flows through the valley.

Across much of the area, gappy hedges enclose rough pastures that are grazed by the skewbald and piebald ponies so characteristic of the Black Country. This untended countryside is a haven for Carrion Crows and Magpies, their depravations doing much to limit the populations of small birds. There are also four golf courses which attract many passage Wheatears.

Within the valley two areas are of special importance. Firstly, the Severn-Trent Water Authority (STWA) has recently constructed a balancing lake and two small islands in a wide meander of the River Tame. Water sports such as sailing occupy most of the lake, but 25 acres (10 ha) at the eastern end have been leased to the RSPB for a nature reserve. The reserve also contains a marsh and some fields and hedgerows — all overlooked by a new Nature Centre. There are also two hides closer to the lake and marsh.

Secondly, there is the huge 1,000 acre (400 ha) expanse of Sandwell Valley Country Park, administered by Sandwell Borough Council. It lies on the opposite side of Forge Lane to the balancing lake and is bisected by the M5. The area west of the motorway is used for formal recreation and is not especially attractive to birds, except for the marshy woodland of Sots Hole. To the east the landscape is surprisingly varied, with woodland, scrub and rough grassland set amidst marginal farmland with neglected hedges. Interest centres on the small Swan Pool, which despite board and dinghy sailing and fishing is still visited by many birds. North of Swan Pool a reclaimed pit-head has been reseeded and planted with saplings.

## Species

Sandwell Valley is an excellent place to watch birds at any time of the year, with always the chance of something unexpected turning up.

On cool, damp days in spring and summer hundreds of Swifts come from their breeding sites in the conurbation and feed low over the balancing lake. A few Swallows from the adjoining farmland join them. Late in summer they might fall prey to a dashing, migrant Hobby. Reed Buntings, Yellowhammers and Whitethroats sing from hedgerows and scrub, with two or three pairs of Lesser Whitethroat where growth is tallest and thickest. A few Yellow Wagtails nest amidst the rough grassland and sometimes a pair of Whinchats or Grasshopper Warblers breed. During the evening Snipe may drum over the damper hollows and later still a roding Woodcock might be seen against the night sky. Willow Warblers are widespread, and areas like Priory Wood hold several Blackcaps, a few Garden Warblers and commoner woodland species. Exceptionally, a Wood Warbler might even be heard, though breeding is unlikely. Around the lakes and pools are one or two pairs of Sedge and Reed Warbler, and a few pairs of Little and Great Crested Grebe, Canada Goose, Mallard and Tufted Duck. Low water levels sometimes expose suitable sites for Little Ringed Plover to nest.

Late summer brings waders to the shorelines. Variety is good, with Oystercatcher, Little Ringed Plover, Ringed Plover, Dunlin, Ruff, Redshank, and Common and Wood Sandpipers all liable to occur, but only in ones or twos. If you are really lucky something like a Pectoral Sandpiper might even appear. Usually there are one or two terns, including perhaps Little or Black, and occasionally a Little Gull. Passerines also move through in great variety, if not large numbers. Wheatear, Whinchat, and later Stonechat are all visitors to grassy areas or scrub, while a Redstart or Pied Flycatcher might be glimpsed working its way along a hedgerow.

Sometimes migrant passerines and waders stay well into autumn and some, like Green Sandpiper, may still be around at Christmas. As summer fades, though, arriving winter visitors become the centre of attention. Soon, good numbers of Snipe are probing the oozy mud or rushy grassland, perhaps accompanied by up to half-a-dozen Jack Snipe and two or three Water Rails. Around the shoreline several Grey Herons watch patiently for fish, while on the water one or two voracious Cormorants might be diving for them. Mixed flocks of Greenfinches, Goldfinches and Linnets gather, while a handful of Corn Buntings may join flocks of Tree Sparrows and Yellowhammers, all searching for seeds. Sometime during winter usually sees small flocks of Siskin and Redpoll. Kestrels regularly hover overhead, especially by the motorway and in late afternoon a Short-eared Owl might search rough grassland for voles.

Wildfowl flock onto the lake. Usually Great Crested Grebes peak first, with a couple of dozen in September, then some 15 Mute Swans gather by mid-autumn. Most species reach their maximum around the turn of the year, when up to 200 Canada Geese, 100 Pochard and Tufted Duck, and 100 Teal can be seen. Small numbers of Wigeon, Goldeneye and Ruddy Duck are often present as well, while Gadwall, Shoveler and Goosander are irregular visitors. Expectations of a scarcer species are always high, and recent years have brought Black-necked Grebe, White-fronted, Barnacle and Brent Geese, Scaup and Common Scoter to the valley. Exceptionally a wintering Redshank or Ruff might remain in some unfrozen spot.

By March, as many Fieldfare and Redwing move north for the summer, thoughts turn to spring. The resident Sparrowhawks begin to soar and display, maybe digressing to see off a wandering Buzzard. Perhaps a skein of Bewick's Swans will head towards their northern breeding grounds. Suddenly the first Meadow Pipits arrive, bringing with them the hope of a passing Merlin. Waders, too, will be trickling through once more. Oystercatcher, Ringed Plover, Dunlin, Sanderling, Turnstone and Green and Common Sandpipers are most likely, but the diminutive Temminck's Stint has been recorded. One or two Arctic Terns might also pass in early May, while among unexpected visitors could be Rock Pipit, White Wagtail, Ring Ouzel or Pied Flycatcher. On one occasion a wayward Marsh Warbler was heard and seen and Quail have also been recorded. Indeed, almost anything could be seen here, just a mere 4 miles (6.5 km) from the centre of Birmingham.

## Timing

Human disturbance is the main factor. Weekends and Bank Holidays are very busy, and this is also when most sailing and windsurfing occurs. The RSPB visitor centre is used by school parties during term time. Bright, sunny mornings are best for bird song. Autumn wader passage is best after a dry summer, when water levels are low. Wildfowl on the balancing lake tend to be more numerous in very cold weather and most vagrants usually arrive after a stormy spell.

## Access

Leave the M5 at Junction 1 onto the A41 Birmingham to West Bromwich road and turn eastwards towards Birmingham. Pass the West Bromwich Albion football ground and take the next turn left into Park Lane. Continue for about 1½ miles (2.5 km) and park in the car park on the left by the old colliery buildings. From here paths can be followed around Swan Pool and the rest of the country park.

For the RSPB reserve, continue along Park Lane (which then becomes Forge Lane). At the end turn right into Newton Road, continue over the railway bridge and take the second right into Hampstead Road. Continue along this road for ⅔ mile (1 km), then turn right into Tanhouse Avenue (immediately before Hampstead School). Then, where the road bends to the right, turn left into the reserve access. The car park is on the right-hand side. From here, walk over the railway bridge to the Nature Centre.

Alternatively the area can be approached from Junction 7 on the M6 Motorway by turning southwards onto the A34 towards Birmingham. At the traffic lights, turn right onto the A4041 towards West Bromwich, left after 1 mile (1.6 km) into Hampstead Road and then follow the same directions as above.

## Calendar

*Resident:* Little Grebe, Great Crested Grebe, Canada Goose, Mallard, Tufted Duck, Sparrowhawk, Kestrel, Snipe, Woodcock, Tawny Owl, Kingfisher and common passerines.

*April–June:* Oystercatcher, Little Ringed Plover, Ringed Plover, Sanderling (scarce), Dunlin, Green Sandpiper, Common Sandpiper, Turnstone (scarce), passage terns, Swift (after April), Swallow, Yellow Wagtail, White Wagtail (scarce), Whinchat, and warblers including Grasshopper, Sedge, Reed and exceptionally Wood. Possible raptor such as Buzzard or Merlin, scarce wader such as Bar-tailed Godwit or Temminck's Stint, or unusual migrant such as Quail or Ring Ouzel.

*July–September:* Grey Heron, Mute Swan, Hobby (scarce), Oystercatcher, Little Ringed Plover, Ringed Plover, Dunlin, Ruff, Redshank, Wood Sandpiper (rare), Common Sandpiper, Little Gull (scarce), terns including perhaps Little or Black, Swift (scarce in September), Swallow, Whinchat, Wheatear and other departing summer visitors including perhaps a Redstart or Pied Flycatcher.

*October–March:* Cormorant (scarce), Grey Heron, Mute Swan, Wigeon, Gadwall, Teal, Shoveler, Pochard, Goldeneye, Goosander (scarce), Ruddy Duck, Water Rail, Jack Snipe (scarce), possible wintering Ruff or Redshank, Green Sandpiper, Short-eared Owl, passage of Meadow Pipits in October and March, Stonechat, Fieldfare, Redwing, mixed flocks of Tree Sparrows and finches, Siskin and Corn Bunting. Maybe Bewick's Swan, rare grebe or storm-driven sea-duck.

# BRIERLEY HILL POOLS

## Habitat
Hemmed in by housing estates and overshadowed by the enormous tip of the former Round Oak steelworks are three large pools owned by the British Waterways Board (BWB). Variously known as Brierley Hill, Pensnett or Fenn Pools, the primary purpose of these pools is to supply water to the Stourbridge Canal. Today two of them are also used for boating and fishing, while the surrounding scrub and rough grassland are used for casual motor-cycling, horse riding and exercising dogs.

## Species
Although very disturbed, this site still manages to hold some good birds, especially in winter.

Few nests are safe from predation or disturbance, so the breeding community is limited. Little and Great Crested Grebes usually succeed in rearing broods, however, and the latter can sometimes be seen carrying their humbug-striped young on their backs. Mute Swans, too, can be seen incubating or tending their cygnets, and Canada Geese occasionally nest as well. The stands of reedmace usually host two or three pairs of Reed Warblers, their rhythmic songs contrasting with those of Sedge Warblers nesting in the nearby scrub. A few pairs of Whitethroat, Willow Tits and an occasional pair of Blackcap also breed.

In early autumn small, acrobatic flocks of Goldfinches flit from stem to stem as they feed on thistle heads. Later on, Redwings gather to strip the berries from the hawthorn scrub, and a few Bramblings may also pass through. Snipe, plus one or two Jack Snipe and Water Rail, are often concealed among the rushes, sedges and reedmace. With them might be a Woodcock. Wildfowl are never numerous, but there are normally up to 50 each of Mallard, Tufted Duck and Pochard. Other species are more erratic, but Wigeon, Shoveler, Teal, Goldeneye and Ruddy Duck could be seen. Spring or autumn may bring one or two waders, a sprinkling of migrants like Wheatear or Redstart, and even a passage tern or Short-eared Owl.

## Timing
The whole area is much used by local people at all times and for all purposes. Disturbance is greatest during school holidays. Very early in the morning is likely to be the quietest time.

## Access
Leave the A461 Dudley to Stourbridge road at the north end of Brierley Hill High Street. Turn westwards into Bank Street and carry straight ahead into Pensnett Road, which then becomes Commonside. Turn right into Blewitt Street and there is a track on the right-hand side between the houses which leads into the pools. There is limited room for parking at the end of this track.

## Calendar

*Resident:* Little Grebe, Great Crested Grebe, Mute Swan, Canada Goose, Mallard, Willow Tit.

*April–June:* Sedge Warbler, Reed Warbler, Whitethroat and Blackcap. Perhaps passage waders, terns, Short-eared Owl, or chats.

*July–September:* Generally quiet, but perhaps passage waders, terns, or chats.

*October–March:* Pochard, Tufted Duck, Water Rail (scarce), Jack Snipe (scarce), Snipe, Woodcock (scarce), Redwing, Brambling (scarce) and Goldfinch (early autumn). Occasionally Wigeon, Teal, Shoveler, Goldeneye and Ruddy Duck. Perhaps Short-eared Owl.

# SALTWELLS WOOD & DOULTON'S CLAYPIT
## Habitat
The 62 acres (25 ha) of Saltwells Wood and Doulton's Claypit are surrounded by houses and the new factories of the Dudley Enterprise Zone. Administered as a Local Nature Reserve (LNR) by Dudley Borough Council, woodland is the dominant habitat, with the 200-year-old trees in Saltwells believed to be relics of the old Pensnett Chase. Pedunculate oak in association with birch is widespread, forming a closed canopy beneath which there is a field layer of bracken, but only a sparse shrub layer with just a few hollies. Within the wood are several old bell pits and two small, but fast-flowing streams. Their valleys are more varied, with beech, ash, lime, grey poplar and sycamore above a richer shrub layer. A stand of alder occupies the south-east corner.

The claypit, quarried last century for fireclay, now exhibits various stages in the natural succession to oak woodland. There are damp patches and wet flushes along the quarry floor, and plenty of birch scrub and developing oak woodland to the south and east. In places gorse is plentiful and there is some open grassland and hawthorn scrub. Two or three adjacent grazing meadows complete the range of habitats. The whole area is used, or abused, by local residents for a variety of purposes.

## Species

For an urban site, this is an excellent place for woodland birds. The sparse shrub layer limits the range of breeding species, but there are good numbers of hole-nesting species like Starling, Blue Tit and Great Tit. Two or three pairs of Nuthatch, Great Spotted and Green Woodpeckers, Stock Dove and Tawny Owl breed in holes and cavities too, while Mistle Thrush, Spotted Flycatcher and Treecreeper also nest in the trees. Blackbirds, Robins, Wrens, Dunnocks and Willow Warblers are widespread and numerous, while Bullfinches, Chiffchaffs and Blackcaps are well represented, with four or five pairs apiece. There are several pairs of Whitethroat and Linnet in the gorse around the claypit, where Jays are especially numerous in the birches and young oaks. Through burying acorns they help to regenerate the oaks. Most unexpectedly for an urban wood, two or three pairs of Wood Warbler breed. Clearly the habitat is very much to their liking, with plenty of low branches for song posts and perches, and an uninterrupted drop to their nests on the ground.

Resident species like woodpeckers and Nuthatch are easier to see when the trees are bare. Autumn can be especially rewarding, with active parties of Redpoll hanging from slender birch twigs, and Redwings feeding on holly and other berries. Less often, a small flock of Siskin visits the alder plantation to feed, or a few Bramblings join with Chaffinches to search for beechmast. One or two Woodcock are also present, cleverly camouflaged in the bracken by day, but emerging to feed in the damper areas at dusk. Even a Grey Wagtail might be seen along the stream.

*Grey Wagtails can be seen at many Black Country sites in the winter*

In the meadows, thrushes, Blackbirds and Starlings seek earthworms and other invertebrates. Green Woodpeckers search for anthills among the rougher areas and small flocks of finches feed on a variety of seeds. Kestrels regularly hover over the open grasslands and a Sparrowhawk might swoop past at any time in pursuit of some small bird.

## Timing

There is considerable disturbance from local residents at all times, but specially during school holidays. Early on a bright spring morning is the best time to visit.

## Access

Approach via the A4036 from Dudley or Stourbridge. Park at the bottom of Merry Hill or in surrounding side roads and follow any of the paths into Saltwells Wood. Continue through the wood to reach Doulton's Claypit.

## Calendar

*Resident:* Sparrowhawk, Kestrel, Stock Dove, Collared Dove, Tawny Owl, Green Woodpecker, Great Spotted Woodpecker, Wren, Dunnock, Robin, Blackbird, Song Thrush, Mistle Thrush, Blue Tit, Great Tit, Nuthatch, Treecreeper, Jay, Magpie, Starling, Chaffinch, Greenfinch, Linnet, Redpoll and Yellowhammer.

*April–June:* Cuckoo, Whitethroat, Blackcap, Wood Warbler, Chiffchaff, Willow Warbler and Spotted Flycatcher.

*July–September:* Very quiet in July as adults tend young. Summer visitors leave in August and September.

*October–March:* Woodcock (scarce), Grey Wagtail (scarce), Redwing, small finch flocks perhaps including Brambling and Siskin.

## HIMLEY & BAGGERIDGE

### Habitat

The parkland of Himley Hall and the adjoining Baggeridge Wood, though on the edge of the Black Country, are never far from urban influence. The central feature of the parkland is the Great Pool. Beyond this is Himley Hall, now an educational centre, and behind that a backcloth of majestic beeches. In the valley above the Great Pool, the feeder stream has been dammed to create four smaller, reed-fringed pools. Originally built to power an ironworks, they were subsequently abandoned and silted up. Recently they have been cleared out again. Higher up is Baggeridge Wood, an oak-birch wood with some interesting birds despite disturbance from motor-cyclists and the like. The whole area is administered as a country park by Dudley Borough Council.

### Species

Although not outstanding for birds, this is a good place for beginners to familiarise themselves with some of the commoner species. Regular watching could well be rewarded with something more interesting.

Rooks, Jackdaws and Stock Doves are often around and Rooks returning to their tree-top nests are an early sign of spring. Overhead, Sparrowhawks display and before long the woods resound to the various calls of Great Spotted Woodpeckers, Green Woodpeckers, Jays and Nuthatches. Soon newly-arrived Chiffchaffs and Willow Warblers add their songs to those of the resident species. In places one or two Wood Warblers can be heard as well. The last migrant to arrive is usually the unobtrusive Spotted Flycatcher, which is seldom seen before early May. On the pools, broods of Mallard, Moorhen and Great Crested and Little Grebes can be seen, while in the reeds a few pairs of Reed Warbler busily tend their young. As night falls a roding Woodcock might show against the night sky.

As summer fades, many breeding birds depart and, once the leaves start to fall, parties of tits and Chaffinches begin searching for beechmast. With them may be a handful of Brambling, their distinctive white rumps identifying them as they take flight. In autumn the pools sometimes hold a Water Rail and once a small party of Bearded Tits even paid a brief visit.

From September onwards small flocks of wildfowl build up on the Great Pool. Most numerous are Mallard, Tufted Duck and Pochard, with the latter sometimes exceeding 100. Some 20 or so Mute Swans may also be seen late in the year. A few Little Grebes are regularly present too, while Teal, Wigeon, Shoveler, Ruddy Duck and even a Goosander might appear. Grey Herons also fish here frequently. Mid-winter could bring something even more unusual, such as a rare diver or grebe. In recent years both Black-throated Diver and Slavonian Grebe have been seen.

In winter the alders along the feeder stream are always worth checking for parties of Goldfinch, Siskin and Redpoll, whilst the weirs are often frequented by Pied and Grey Wagtails.

## Timing

Himley Park is extremely popular at weekends and evenings during the summer, so an early visit is recommended. During the winter, weekend sailing may disturb wildfowl and in very cold weather the pools ice over completely. On the whole the woods are quieter, apart from motor-cyclists.

## Access

From the A449 Wolverhampton to Kidderminster road turn eastwards at the traffic lights by Himley Church and take the B4176 towards Dudley. The entrance to Himley Hall is on the left in about ½ mile (0.8 km). There is plenty of car parking (a charge is made) and obvious paths lead around the lake and up past the millpools into the wood.

## Calendar

*Resident:* Little Grebe, Great Crested Grebe, Mallard, Sparrowhawk, Moorhen, Woodcock, Stock Dove, Green Woodpecker, Great Spotted Woodpecker, Nuthatch, Rook, Jay and other common woodland birds.

*April–June:* Common migrants including Reed and Wood Warblers and Spotted Flycatcher.

*July–August:* Generally busy with people and quiet for birds as summer visitors leave.

*September–March:* Grey Heron, Mute Swan, Water Rail (scarce), Grey Wagtail and parties of tits and finches including Goldfinch, Siskin, Redpoll and maybe Brambling. Perhaps Wigeon, Teal, Shoveler, Ruddy Duck or Goosander. Possible diver or rare grebe.

Birmingham is Britain's second city and the economic and social capital of the West Midlands. A bustling, busy mass of concrete, bricks and tarmac, it is unlikely to be visited by anyone in search of birds. Yet for those who find themselves with time to spare in this great city, some surprisingly pleasant places and unexpected birds can be found.

## Habitat

There is no great river, but the canals make up for this shortcoming as they weave their way into the very heart of the city. Before the railways and motorways, they were the arteries of this great industrial empire. Now the traditional industries have gone and the canals stand silent and neglected. Most are still navigable, however, and some are beginning to enjoy a welcome resurgence as cruising waterways. At the same time all are steadily being reclaimed by nature and this is bringing wildlife back into the very centre of the city. The small rivers, too, are being steadily cleaned up and, though the Rea has probably been too extensively canalised to ever again attract much wildlife, stretches of the Cole are much better.

Industrial decline during the 1980s has added to the legacy of derelict land and this also is being swiftly recolonised by wildlife. Some sites, like the sidings at the old Snow Hill station, even lay derelict long enough to sprout sturdy saplings.

Most urban parks and open spaces are too tidily managed to be really good for birds. There are two exceptions, namely the steep, wooded hillsides of the Lickeys, and the heaths, woods and pools of Sutton Park. Overall the water-supply and canal-feeder reservoirs, some in pleasant countryside just beyond the reach of bricks and mortar, are of most interest.

## Species

Pigeons are most obvious and Starlings most spectacular. Noisy hordes of the latter are a familiar sight huddled together along the ledges of city centre buildings or in churchyard trees. Over 50,000 roost here on some winter nights, the highest numbers not necessarily coinciding with the coldest weather.

During the daytime, small flocks of Goldfinches or Linnets, accompanied by the occasional Reed Bunting, feed on derelict sites. Migrant Wheatear and Yellow Wagtail have also been seen, and even Red-legged Partridge and Corn Bunting have appeared in such unlikely spots. Kestrels and even Tawny Owls are attracted by the many rodents in these areas. Sparrowhawks, too, can now be seen quite close to the city centre.

The real urban speciality, though, is the Black Redstart. This unobtrusive little bird haunts large derelict buildings and structures in power stations, gas works, railway stations and factories, especially those close to water. It is best located from canal towpaths in the early morning, particularly Sundays, when background noise from traffic is least. Then the sooty black males can be heard singing from the highest roofs and chimneys. Precise localities are hard to identify, as new development is constantly changing the scene. The old Snow Hill Station is a traditional haunt, but as this is now being redeveloped birds may move elsewhere. However, the 3 mile (5 km) stretch of the Birmingham and Fazeley Canal from Gas Street Basin to Spaghetti Junction should prove productive. Breeding Black Redstarts arrive in late April or May and leave again in August, although one or two might still be present in September.

*Black Redstarts breed in the centre of Birmingham*

Jays, Magpies and Crows occur in sylvan suburban gardens and in woods, where they are joined by other woodland species such as warblers and woodpeckers. Even this far inland, gulls are numerous and parties can be seen standing around on school playing fields and recreation grounds. Most are Black-headed, but careful scrutiny might reveal a rarity like Glaucous Gull. The more interesting locations are described below.

## Edgbaston Reservoir

### Habitat
Just 1½ miles (2.5 km) from the centre of Birmingham is Edgbaston Reservoir — a canal-feeder reservoir covering 64 acres (26 ha) with gently shelving, gravelly shores and a few marshy creeks. Hemmed in on three sides by housing and on the fourth by old factories, its setting is truly urban, but the variety of mature trees and herbaceous plants is better than in more formally managed parks. Of interest to botanists are amphibious bistort and a good stand of flowering rush. Despite disturbance from fishing, boating, sailing and general public access, regular watching has produced a good variety of birds. The area is administered by Birmingham City Council.

### Species
It would be unwise to have great expectations of this site, which is mentioned principally because it is so close to the city centre.

Small numbers of Great Crested Grebe, Canada Geese, Mallard, Tufted Duck and Pochard are winter regulars. Occasionally they are joined by one or two Wigeon, Teal, Gadwall or Goldeneye. On a good night there are up to 6,000 roosting gulls as well. Most are Black-headed, with a scattering of Common, Lesser Black-backed and Herring Gulls with them. Two or three Great Black-backed Gulls or a Kittiwake may add to the variety, or exceptionally a Glaucous or Iceland Gull. Surprisingly the region's first ever Laughing Gull was seen here recently. The list of vagrant wildfowl is also impressive and includes Great Northern Diver, Red-necked Grebe, Ferruginous Duck, Scaup and Long-tailed Duck.

Small numbers of terns are noted on passage, most often in spring, with Arctic and Common most regular and Black Terns most erratic. Even the rare White-winged Black Tern has made an appearance. Wader records are few, but an occasional bird does stop by to feed briefly or rest.

Interest is not wholly centred on the water. Grey Wagtails sometimes breed and small numbers of the commoner migrants like Meadow Pipit and Willow Warbler regularly pass through. In autumn, Linnets may gather to pick up seeds from any exposed mud, or a Rock Pipit could work steadily around the shoreline. Sometimes a Chaffinch flock is joined by one or two Brambling. With the passage birds may come something unusual. In recent years, Merlin, Hobby, White Wagtail and incredibly Woodlark have all helped to demonstrate the range of migrating birds that might be seen passing over a major city.

## Timing

The reservoir is a popular and well-used recreational facility at all times. Nevertheless it repays regular visits, particularly during the winter. It is quietest on weekdays outside the school holidays, when it is the ideal place to while away a lunch-hour. Often it is most productive after a dry spell, when the water level is low.

## Access

The best approach is from the Hagley Road (A456 Birmingham to Kidderminster). At the Ivy Bush public house, about ½ mile (0.8 km) west of the Five Ways underpass, turn northwards into Monument Road. Then take the fifth turn left into Reservoir Road and continue to the end. Park inside the reservoir, the entrance to which is at the end of this road, and then follow the paths around.

## Calendar

*Resident:* Common wildfowl and passerines, including perhaps Grey Wagtail.

*March–June:* Meadow Pipit in March. From April possible White Wagtail, warblers and maybe terns including Black.

*July–September:* Parties of Linnets and other finches. Perhaps a migrant Hobby or terns.

*October–February:* Great Crested Grebe, Canada Goose, Wigeon, Gadwall, Teal, Mallard, Pochard, Tufted Duck, Goldeneye, roosting gulls including perhaps Iceland and Glaucous (January–March) or Kittiwake, and parties of finches possibly including a few Brambling. Perhaps a diver, scarce grebe or storm-driven sea-duck. Maybe a passing raptor or Rock Pipit in autumn.

# BARTLEY RESERVOIR

## Habitat

Bartley lies just over 5 miles (8 km) south-west of the city centre. For many years this 114 acre (46 ha) water-supply reservoir was 'Mecca' for Birmingham's birdwatchers. Now, through his book *Gone Birding*, Bill Oddie has immortalised it as a place totally devoid of birds! All too often this is true, but regular watching still reaps a rich harvest of species.

Once deep in open country and reached through a maze of narrow lanes, Bartley is now overshadowed by the tower blocks of Birmingham. With solid concrete banks and short grass swards fortified by a chain-link perimeter fence, it is bleak and uninviting. As the water in the reservoir has previously been partially treated, the Severn-Trent Water Authority (STWA) is very sensitive about possible pollution, so elaborate bird-scarers are used to dissuade gulls from roosting. As the records testify, however, their success has been questionable. For the same reason recreational activity was originally resisted, but recently both sailing and windsurfing have been introduced.

To the west of the reservoir is Birmingham City Council's Bromwich Wood, a small oakwood known locally as Bluebell Wood because it is carpeted with these spring flowers. In addition to the oaks, there are ash, sweet chestnut, wild cherry, alder and rowan above an understorey that abounds with autumn berries on hawthorn, elder, holly, rose, honey-suckle and bramble. This larder is sufficient to sustain fair numbers of resident and migrant birds, while typical farmland species can be found in the adjoining fields and hedgerows.

## Species

Winter wildfowl and gulls are the main interest, but Bartley can be agonisingly erratic. Often a couple of Goldeneye are all that are present during the day. Even the gull roost is unpredictable, with several thousand birds one night and none the next.

In mild weather, up to 50 Tufted Duck (exceptionally 100), one or two Great Crested Grebes and Coot, and perhaps a dozen or so Mallard and Pochard are typical. Wigeon are less regular, but small numbers are sometimes present in March. Occasionally a larger flock of 100 or more arrives earlier in the winter and some may then remain until the spring. Shelduck, Pintail, Shoveler and Goosander are only irregular visitors in ones or twos.

Hard weather is usually more exciting, as Bartley is deep and therefore freezes over later than most waters. As a result it acts as a refuge for wildfowl from miles around. During a prolonged freeze, a diver, one of the rarer grebes or a sea-duck is quite likely to appear and perhaps stay for several days or even weeks. Recent years have seen Great Northern Diver, Bewick's and Whooper Swans and White-fronted Geese arrive in such circumstances. Of the sea-duck, Common Scoter is almost annual, with spring, late-summer and autumn occurrences associated with regular overland migrations. Drakes predominate in the first two of these and ducks and immatures in the latter. The coldest nights also attract the most roosting gulls, with up to 4,500 Black-headed, 1,500 Lesser Black-backed and 150 Herring Gulls. With them may be a Glaucous or Iceland Gull, one or other of which is reported most years.

Once the wintering wildfowl and gulls have left, the reservoir is largely deserted, although Mallard breed. A few waders and terns are noted each spring and autumn, staying only for a short time if at all. This is hardly surprising, since even with a low water level the shoreline is suited only to such birds as Common Sandpiper or Rock Pipit. Nevertheless, very small numbers of Oystercatcher, Ringed Plover or Dunlin might pause briefly in spring. This is also the best season for terns, with Arctic and Common recorded regularly and Black Tern occasionally. Sometimes terns are accompanied by a Little Gull. Autumn passage frequently produces fewer species, but Dunlin, Ruff, Greenshank and Common Sandpiper are fairly regular.

By October wildfowl are gathering again and parties of Meadow Pipits and Skylarks move westwards, particularly on misty mornings. This is the most likely time to find a Rock Pipit or a rarity such as Snow Bunting or Short-eared Owl.

Bromwich Wood is often worth a quick look. Residents include thrushes, tits, Jay, Nuthatch, Treecreeper and Great Spotted Woodpecker. In spring they are joined by warblers and Spotted Flycatcher, while in autumn passing migrants might exceptionally include a Firecrest. In late autumn and winter mixed finch and bunting flocks gather in adjoining stubble, and Fieldfares and Redwings strip hips and haws from thorn hedges, before turning to pastures in search of earthworms.

## Timing

Water sports mostly take place at weekends, while bird-scaring in the winter seems to be erratic. Southerly winds are usually best for bringing migrants in spring, while misty mornings are best in autumn. In winter, the reservoir holds most birds in very cold weather, when other waters are frozen solid. Periods during or immediately after storms are quite likely to bring vagrant wildfowl.

## Access

The reservoir is best approached from the A38 Birmingham to Worcester road. Turn northwards at the traffic lights by the Bell Inn at Northfield and carry on down Bell Hill and into Shenley Lane. Continue over the next hill and then turn left at the foot into Long Nuke Road. Carry on to the end of this road and then either turn right to view the reservoir from the road across the dam, or better still turn left and then second right to follow Frankley Lane around the south side of the water. Views can be obtained from this lane, but it is very narrow and parking is difficult. It is better to continue to the next junction and turn right into Scotland Lane. View the reservoir from anywhere along this lane, where parking is easy. Bromwich Wood is on the opposite side of Scotland Lane to the reservoir and there are obvious access points and tracks around.

## Calendar

*Resident:* Mallard, Great Spotted Woodpecker and common passerines including Nuthatch, Treecreeper and Jay.

*April–June:* Oystercatcher, Ringed Plover, Sanderling (scarce), Dunlin, Common Sandpiper, perhaps Little Gull or terns including maybe Black, common woodland warblers and Spotted Flycatcher. Maybe Common Scoter.

*July–September:* Dunlin, Ruff, Greenshank, Common Sandpiper, perhaps terns including maybe Black. Maybe Common Scoter. Wildfowl begin to gather in September.

*October–March:* Great Crested Grebe, Shelduck (scarce), Wigeon (mostly March), Pintail (rare), Shoveler (scarce), Pochard, Tufted Duck, Common Scoter (scarce), Goldeneye, Goosander (scarce), Coot, roosting gulls often including Iceland or Glaucous (January–March), Skylarks and Meadow Pipits (October and November), Rock Pipit (October and November: rare), Fieldfare, Redwing and mixed finch and bunting flocks. Perhaps a diver, rare grebe, wild swan or vagrant sea-duck. Possible rarity especially in October and November.

# BITTELL RESERVOIRS

## Habitat

Bittell Reservoirs nestle into the folds of some extremely pleasant countryside just 8 miles (13 km) from the centre of Birmingham and still within sight of the giant Longbridge car plant. Strictly there are only two reservoirs, Upper Bittell which covers 100 acres (40 ha) and Lower Bittell which covers 57 acres (23 ha), but the latter is divided into two by a road across its northern corner. Between the two reservoirs a tiny, fast-flowing stream winds its way through a productive patch of damp, mixed woodland. A feeder arm from Upper Bittell passes east of Lower Bittell to join the canal. British Waterways control access to the reservoirs.

Excluding dams, both reservoirs have gently shelving, natural shorelines backed by rough grassland where wildfowl graze. After a summer of heavy canal traffic, the low water level also exposes muddy feeding areas for waders. Both reservoirs are fished and there is sailing on Upper Bittell, but no public access except on roads and footpaths, so disturbance is not too great. Lower Bittell is quieter, more sheltered, and often preferred by wildfowl despite its smaller size.

## Species

At times there is a considerable movement of birds between Bittell and Bartley, but Bittell usually carries the greater number and variety. Again winter wildfowl and gulls are the main attraction.

The commoner wildfowl often peak in late summer or early autumn, when there are typically 300 or so Canada Geese, almost as many Mallard, 200 Coot and well over 100 Tufted Duck. Flocks, especially of geese, then decline during the autumn, but Mallard and Tufted Duck often peak again in mid-winter. From late autumn to early spring up to 100 Teal and 50 Pochard can be seen. Small numbers of Wigeon, Shoveler, Ruddy Duck and Great Crested Grebe are also present throughout the winter, with the latter peaking at around 30 each spring. A few Goldeneye usually arrive in November and stay until the following March, but Shelduck, Gadwall and Goosander are irregular visitors.

An intriguing phenomenon concerns Barnacle Geese. A flock of 29 first arrived in December 1983 and remained until the following February. At the time they were thought to be feral birds, an opinion reinforced when three reappeared the following August. However, a further 29 then arrived in November and this pattern has been repeated ever since, with the flock increasing to over 70. It seems just possible that a small nucleus of feral birds has acted as a decoy and pulled in some genuinely wild stock. Other recent rarities have included Black-throated and Great Northern Divers, Red-necked and Slavonian Grebes, Shag, Scaup, Common Scoter, Smew and Red-breasted Merganser.

The gull roost is not large, with typically only 1,500 Black-headed and 200 or so Lesser Black-backed and Herring Gulls, but twice these numbers assemble on occasions. Despite being a small roost there is still a good chance of an Iceland Gull, which in recent years has been commoner here than Glaucous Gull.

Up to a dozen Grey Herons stand motionless in the shallows, while probing Snipe emerge from behind clumps of rush and sedge. With them may be one or two Jack Snipe, but they are unlikely to be seen unless flushed. Water Rails are equally secretive, and the most likely indication of their presence will be loud squeals coming from lush vegetation.

As the wildfowl and gulls depart at the onset of spring, thoughts turn to waders, terns and summer migrants. Early passage may bring a Kittiwake, Little Gull or Oystercatcher in March, but it is April and May before movement really peaks. Waders in spring are few and their stay brief, but Little Ringed Plover, Redshank and Common and Green Sandpipers are seen most years, with perhaps Ringed Plover, Dunlin and Sanderling too. Terns seldom appear in numbers either, but there is always the chance of a larger Sandwich Tern as well as the more usual 'Commics'. Recently a Night Heron made a brief spring visit.

Most commoner summer migrants pass through or stay to nest. Along the lane to Upper Bittell, Willow Warbler, Chiffchaff, Blackcap and Garden Warbler should be singing in the woods, Sedge and Reed Warblers in the willow scrub and reed-beds, and Lesser Whitethroats in thick hedgerows. Among the more interesting residents, Bullfinches breed in thick hedges and scrub, Willow Tits nest in rotting stumps, all three species of woodpecker can be found in the woodland, and Kingfishers usually raise a brood or two somewhere around the reservoir or along the canal feeder. Breeding wildfowl are not outstanding, but a few Little Grebe, Mallard and Coot nest regularly and Great Crested Grebe attempts to breed from time-to-time.

The low autumn water levels usually attract more waders. Passage begins in July, when Green and Common Sandpipers may reach their peak, and continues through August and September, when Little Ringed Plover, Ringed Plover, Dunlin, Ruff, Redshank and Greenshank are the most likely species to be seen. Numbers are small, seldom more than ten, but almost anything might appear, with recent records of Temminck's Stint, Purple Sandpiper, Little Stint, Black-tailed Godwit, Curlew and Wood Sandpipers, and Spotted Redshank. Later in the autumn a solitary Knot, Grey Plover or Bar-tailed Godwit may make a brief visit. Tern numbers are again small, but there is still a chance of a Little, Sandwich or Black Tern. The late-summer flocks of wildfowl might also contain a Garganey, and this is the time of year when Hobbies are quite often seen and both Osprey and Peregrine have appeared. Autumn storms sometimes bring a vagrant seabird such as an exhausted Shag or an Artic Skua.

Although the wildfowl and gulls inevitably command attention, the variety of common passerines is good in winter too. Chaffinches, Greenfinches and Yellowhammers congregate to feed in stubble and plough, and sometimes a few Brambling join them in late autumn or early spring. In hard weather birds frequently gather around cattle and sheep troughs. The large alders alongside the path from Lower to Upper Bittell are a regular haunt for flocks of Redpoll and Siskin, the latter often more than 100 strong. Parties of Bullfinches and Long-tailed Tits frequent this lane as well, while the sheltered valley is a favoured spot for Redwings and Fieldfares in late autumn. One or two Grey Wagtails usually winter along the stream, Kingfishers periodically fly along the valley and even a Dipper has been seen beneath the waterfall.

## Timing

There are usually birds to be seen at any time of day or year. Weekends are busiest and that is when most sailing and fishing occurs. Otherwise weather is a key factor. In spring the best birds usually arrive in southerly airstreams with overcast skies, while a dry summer is often followed by a good wader passage. Autumn gales may bring something unusual, but winter wildfowl may be driven out by an extremely cold spell, when both reservoirs can freeze.

## Access

Approach from the B4120 Alvechurch to Lickey Road, which skirts the southern shore of Lower Bittell. There are good views from this road, but parking is difficult as it is narrow and there are double yellow lines around the bends. It is better to turn northwards into the lane towards Hopwood, park on the verge and walk back to the main road, where the tall hedge is usually cut in one or two places to permit easy observation. Alternatively continue towards Hopwood and park on the verge where the lane bends sharp right to pass across the upper end of the reservoir. Good views can sometimes be obtained from here, but it is looking southwards so beware of the position of the sun.

For Upper Bittell return to the bend and proceed on foot up the track alongside the reservoir. Do not take cars along as it is a very rough, private track with no adequate parking. Moreover, the walk along here should never be missed or rushed as it is often rewarded with something of interest. Follow the track through the wood and round a sharp right bend by a small pond. Then cross the style in front of you and climb the footpath to view the reservoir from the top of the dam. Return to the main track and turn right past the sailing club. At the end turn right again and follow the North Worcestershire Path past Cofton Richards Farm. This eventually brings you to the northern shore of the reservoir.

## Calendar

*Resident:* Little Grebe, Great Crested Grebe, Mallard, Sparrowhawk, Coot, Kingfisher and common woodland birds including all three woodpeckers, Willow Tit and Bullfinch.

*March–June:* Winter visitors leave during March and April. Oyster-catcher, Little Ringed Plover, Ringed Plover, Sanderling (scarce), Dunlin, Redshank, Green Sandpiper and Common Sandpiper. Perhaps Little Gull, Kittiwake or terns including the occasional Sandwich. From

April, Sedge Warbler, Reed Warbler and common scrub and woodland warblers. Perhaps Shelduck, or a rarity in May or June.

*July–September:* Grey Heron, common wildfowl, perhaps Shelduck or Garganey, Little Ringed Plover, Ringed Plover, Dunlin, Ruff, Redshank, Greenshank, Green Sandpiper and Common Sandpiper. Less often stints, Curlew Sandpiper, godwits or Wood Sandpiper. Terns including occasional Sandwich, Little or Black. Perhaps a migrant raptor, storm-driven seabird or other rarity in September.

*October–February:* Grey Heron, Canada Goose, Barnacle Goose, Shelduck (scarce), Wigeon, Gadwall (scarce), Teal, Shoveler, Pochard, Tufted Duck, Goldeneye, Goosander (scarce), Ruddy Duck, Smew and Red-breasted Merganser (rare), Water Rail, Jack Snipe (scarce), Snipe, occasional late wader such as Grey Plover, roosting gulls maybe including Iceland (scarce) or Glaucous (rare) after December, Grey Wagtail, Fieldfare, Redwing, mixed finch and bunting flocks, perhaps including Brambling, Siskin and Redpoll. Maybe a diver, rare grebe or storm-driven seabird.

# LICKEY HILLS

## Habitat

Barely 1 mile (1.6 km) to the west of Bittell Reservoirs and right on the city boundary, the Lickey Hills rise like a knife edge to a height of 956 ft (291 m). The hills are managed by Birmingham City Council and there is free public access. From the summits of this Cambrian and Silurian ridge, there are extensive, panoramic views across pleasant, undulating countryside to the south and west, and across Birmingham to the north and east.

Some summits and hillsides, such as Beacon Hill, are open grasslands, but the majority are clothed with plantations of pine and larch interspersed with native broad-leaves among which oak, beech, birch and sycamore are all locally dominant. On the steeper, more open hillsides patches of bilberry, gorse and heather are reminders of the indigenous vegetation. Beneath the pines and beeches are nothing but needles and fallen leaves, but along the valleys the more open oak canopy has a rich understorey of hazel, holly, rowan and bramble. There is plenty of birch scrub too, especially among the wetter areas.

The older pinewoods are mostly along the main ridge from Cofton Hill, with the larch plantations on the lower slopes to the west. The best broad-leaved woodland is in the valley that runs down from Twatling Road and at Sunnybank on the opposite side of Rose Hill.

The hills are popular for walks, picnics, horse riding, exercising dogs and many other recreational pursuits. If there is snow lying, Beacon Hill is used for skiing and sledging. Despite all this, the birdlife remains good.

## Species

Few rare or unusual birds are likely to be seen here, but a good range of typical woodland species makes this an ideal place for beginners to familiarise themselves with songs, calls and general jizz. In winter it is one of the most reliable places for Brambling.

Common residents like Wren, Dunnock, Robin, Song Thrush, Blackbird, Blue and Great Tits, and Chaffinch are widespread and abundant. All are easiest to locate once they have started to sing in early spring. This is also a good time to seek out the less common residents. Treecreepers are quite numerous and widespread, being equally at home among the conifers and broad-leaved trees. Yet they are seldom obvious. As a rule they prefer the smaller trees and saplings, where they nest in the nooks and crannies behind ivy or peeling bark. Though less widespread, Nuthatches are more obvious, especially in spring when their shrill calls carry far through the woods. They are closely associated with the mature stands of oak and beech, where they are quite common.

Woodpeckers are not as rare or restricted as is sometimes thought. Green Woodpeckers are just as likely to be seen searching out anthills amongst the short hillside turf as they are in the woods. Usually you hear them first, then catch sight as they bound away in undulating flight. Great Spotted Woodpeckers are much more arboreal and harder to track down if you are unfamiliar with their explosive 'tchick' calls. In spring, though, their sharp, resonant drumming echoes across the valleys. They too favour the mature stands of oak and beech, though occasionally one is seen amongst the conifers. Rotting birch stumps are good for nest-holes. Hardest of the three woodpeckers to find is the Lesser Spotted. Not only is it the smallest and scarcest, but it chooses to flit around the topmost branches of oaks and beeches. Again it is best located by sound, either through its shrill 'pee-pee-pee' call or its weaker drumming. Among the other residents, the noisy Jays, Magpies and Carrion Crows make their presence only too apparent.

Summer visitors are plentiful on the hills. Chiffchaffs and Willow Warblers return early to their territories and can be watched darting after insects among the bare branches. Chiffchaffs again often prefer the mature oakwoods, where they frequently sing from the highest branches. Blackcaps and Garden Warblers, on the other hand, keep more to the shrub layer, which by the time they return to their territories is often in full leaf. With luck, some scarcer species might even be encountered. A few pairs of Wood Warbler are spread thinly across the hills, showing a preference for beech and oakwoods with little undergrowth, Redstarts probably still nest here and there, though their numbers have fallen of late, and Tree Pipits occur sparingly on open hillsides with scattered trees.

There are plenty of Coat Tits and Goldcrests, their high-pitched calls and thin songs a familiar sound in the conifer plantations. The trills of Redpoll can also be heard. If you do stumble across a party of Goldcrests, check carefully for a Firecrest among them. Firecrests bred in larch trees on the hills in 1975, but have not done so subsequently. More reliable in their visits are Crossbills. In irruption years, small twittering flocks may arrive in the larch and pinewoods at any time from mid-July onwards, the bright crimson males a colourful sight as they cling to a slender spray and strip the cones as they feed. Sometimes they pass on again quite quickly, but another time they will decide to stay for the winter.

The beechwoods, particularly those adjoining Twatling Road, are a winter haven for many birds, especially when beechmast is plentiful. From October onwards large flocks of Chaffinches and Bramblings peck amongst the leaf litter for mast. Chaffinches regularly number up to 300, but Brambling are much more erratic. In some years there are only a couple, but in others 100 or so are by no means unusual and on occasions as many as 400 may gather. Feeding with these finch flocks are often a handful of Great Tits and one or two Marsh Tits. Lower down the valley, parties of tits roam through the birchwoods, acrobatically taking seeds from the ends of slender twigs. With them, in the streamside alders and willows, may be a party of Long-tailed Tits, a small flock of Redpoll or a few Siskin. Just occasionally a couple of Treecreepers or a Lesser Spotted Woodpecker will attach themselves to one of these flocks.

As spring approaches, the resident Sparrowhawks become more obvious as they begin to soar and display above the trees, and a Buzzard might sail lazily past. Sometimes a Pied Flycatcher passes through, or exceptionally a pair arrives to raise hopes of another sporadic breeding attempt. In autumn a migrant Ring Ouzel might visit the hills, probably when there are plenty of rowan berries. Rarities are few, but Golden Oriole has been seen in the past.

## Timing

Spring is the best time to visit. For resident birds, go before the trees are in leaf, but for summer visitors it is better to wait until May, when their numbers are at full strength. Whenever you go you will need to be early in the morning before the locals have begun to exercise their dogs. Weekdays are generally the quietest, although early on a Sunday morning can also be quite good. The hills are best avoided on sunny weekend afternoons, especially in high summer when there are crowds of people and most breeding birds become very secretive. In winter, timing is less critical, though it is still surprising how many people there are around.

## Access

Approach from the A38 Birmingham to Worcester road and, at the roundabout outside the Longbridge car plant, take Lickey Road (B4096) to the south. Continue beyond the end of the dual carriageway and turn right at the roundabout into the Old Birmingham Road. Part way up the hill, on the right-hand side, is a car park and Visitor Centre, from where the hills can be explored to north and south on a variety of tracks. Alternatively, carry on up the hill (known locally as Rose Hill) and turn left at the crossroads into Twatling Road. You can then turn left along a track which leads to another parking area, or you can follow the road to the right and park on the left-hand verge. From either point you can then walk southwards through the beech and oakwoods, following a variety of paths.

## Calendar

*Resident:* Common woodland birds including Sparrowhawk, Kestrel, Stock Dove, Tawny Owl, all three woodpeckers, Goldcrest, Marsh and Willow Tits, Nuthatch, Treecreeper, Jay and Redpoll.

*April–June:* Last winter visitors leave in April. Turtle Dove, Cuckoo, Tree Pipit, Redstart (scarce), woodland warblers including Wood Warbler, and Spotted Flycatcher. Possible Buzzard, Pied Flycatcher or other scarce migrant. Exceptionally Firecrest.

*July–September:* Quiet in July. Summer visitors leave in August. Possible Crossbills in irruption years. Maybe unexpected migrant like Ring Ouzel in September.

*October–March:* Flocks of tits and finches including Brambling and Siskin. Maybe Buzzard or party of Crossbills.

# SUTTON PARK

## Habitat

This incomparable tract of 2,500 acres (1,000 ha) of heath and wood, entirely surrounded by urban development, is one of the largest and finest urban parks in Europe. It has survived because it was given firstly by Henry VIII to the Bishop of Exeter, and secondly by the bishop to the Borough of Sutton Coldfield for the benefit of its inhabitants. The park is now run by Birmingham City Council. Some of the housing that adjoins the park is amongst the most exclusive in the Midlands and the secluded gardens are effectively an extension of the park itself.

The natural vegetation is a typical mosaic of the heath, bog, marsh and wood that must have covered much of the West Midlands before the Industrial Revolution. Across much of the park the poor, acidic soils support only heather, western gorse or heath grasses like wavy hair-grass and mat-grass. In the wetter hollows and valleys like Longmoor, cross-leaved heath, purple moor-grass and cotton-grass take over, with sphagnum mosses in the boggier spots. These wet places form a very important habitat for all forms of wildlife.

Oak is dominant in the older woods, but there is a lot of birch and mountain ash, and holly is a feature of the understorey. The older oaks harbour myriads of insects, rotting birches provide a wealth of nest-sites, and rowan and holly a rich harvest of autumn berries. Parts of the higher, drier ground are covered by bracken or grass heath, with scattered clumps of birch forming an open woodland. In recent years, birch scrub has been steadily invading the heath grassland. Along the streams and around the pools, particularly Bracebridge, Blackroot and Wyndley Pools, the woodland is wetter and dominated by tall alders.

## Species

The main interest is passage and breeding passerines. Movement begins in March with parties of Meadow Pipits and perhaps one or two Stonechats. Up to a dozen Wheatears follow in April, while in May a migrant Hobby may glide across the heath in pursuit of insects. Skylarks and Meadow Pipits are both common breeding species on the grassy heaths and just occasionally a non-breeding Curlew is seen in summer. Some 30 pairs of Yellowhammer and Linnet nest amongst the gorse and heather, but Stonechats breed only spasmodically, usually after a succession of mild winters when their numbers are high. Whitethroats also nest amongst the gorse or in scattered hawthorns, but in fewer numbers, with seldom more than half-a-dozen pairs. Scan the sweeps of bracken for a Whinchat perched on a frond, as one or two pairs nest most years.

After the breeding season, Skylarks, pipits and chats pass through on their return journey south, with Stonechats in some years staying well into autumn. Check any areas that have been burnt by summer fires, as they often attract Wheatears, Whinchats and Stonechats. Once the summer visitors have gone, the open heath is largely empty save for the small resident populations of Pheasant and Grey and Red-legged Partridge. Kestrels and Sparrowhawks are frequently seen at all times and in all seasons, while a wintering Great Grey Shrike, Hen Harrier or Merlin are all possibilities, though none has occurred in recent years.

On still, summer evenings two or three displaying Snipe plunge earthwards with outspread tails over marshy, boggy valleys like Longmoor. Overhead, the roding flights of one or two Woodcock take them round above the tree-tops, while from deep within the lush valley vegetation comes the ventriloquial 'churring' of a Grasshopper Warbler. Once as many as six could be heard reeling, but numbers are currently at a low ebb. Winter brings more Snipe, possibly a Jack Snipe and up to five Water Rails to the marshy valleys and pool margins, while in the twilight as many as a dozen Woodcock may emerge from the bracken-clad slopes and come to the water's edge to feed.

The range of woodland species is good. Commonest are Willow Warbler and Chaffinch, with some 150 and 75 pairs respectively spread across the park. Magpies, Jays and Carrion Crows are also numerous and widespread. Where the open heath merges into light birch woodland, the musical song of a Tree Pipit may be heard. Never very conspicuous, only three or four pairs nest at best and this is another species that seems to be in decline. In the wetter ground, Willow Tits nest in rotting birch stumps, but the sibling Marsh Tit is unlikely to be found far from mature trees. Three or four pairs of each nest most years. In winter foraging parties of tits hang precariously from the ends of delicate birch twigs to take the seeds. With them in early autumn are flocks of Redpoll and Siskin, though these soon move to the mature alders for the rest of the winter. Redpoll are seldom numerous, with 50 to 100 about average, but Siskin are much more variable. A poor year may see only a couple of dozen birds, while a good one might bring 100 or 200.

A few Redpoll also breed, mainly in the conifer plantations, where Coal Tits and Goldcrests are also plentiful. In winter, the commoner pigeons, thrushes and finches roost in these plantations. With them are up to 50 Brambling, with exceptionally five or six times as many. Overall, though, the conifers are dull compared with the broad-leaved woodland.

Some 30 pairs of Chiffchaff and almost as many Blackcaps are spread throughout the older deciduous woodland, all three woodpeckers are well represented and there are a dozen or so pairs of Treecreeper and Nuthatch. By comparison Garden Warblers and Spotted Flycatchers are relatively scarce, while as many as six Wood Warblers can be heard in song, but breeding seldom occurs. Up to ten pairs of Redstart have been known, but this is yet another declining species. None the less, it is a delight to see this handsome bird flitting through the trees.

Around the many pools in summer a pair or two of Reed and Sedge Warblers nest regularly in the reeds and sallows, while Grey Wagtail has bred on occasions too. Of the wildfowl, a few pairs of Great Crested Grebe, Mute Swan, Canada Goose, Mallard and Tufted Duck all nest, while Little Grebes sometimes attempt to do so. Winter wildfowl regularly include small numbers of Great Crested Grebe, Mute Swan, Canada Goose, Mallard and Tufted Duck, augmented from time-to-time by the occasional Wigeon, Teal, Shoveler, Pochard, Goldeneye,

Goosander or Ruddy Duck. Less expectedly, a passing Shelduck or Common Scoter may drop in for a short stay and even a Great Northern Diver has been known to pause awhile. One or two waders, gulls and terns are also seen, with recent records of Oystercatcher, Redshank, Little Gull, Kittiwake, and Common, Arctic and Black Terns.

Sutton Park is not noted for rarities, but Quail, Firecrest and the rich, flutey song of a Golden Oriole have all rewarded the regular watchers in recent years.

## Timing

Sutton Park is a very popular spot for all kinds of casual and organised recreation. Local people use it for walking or exercising their dogs, children use it as a playground and on fine days in summer crowds flock to the park to picnic, ramble or play games. Boating and fishing are also popular pastimes. With so much activity, it pays to arrive early in the morning before everyone else. Fortunately the birds will be most active then as well. In summer, a later evening visit to Longmoor or one of the other valleys can also prove worthwhile. Timing matters less in autumn and winter, but avoid the RAC Lombard rally in November.

## Access

Sutton Park lies ½ mile (0.8 km) to the west of Sutton Coldfield town centre (which is on the A453 Birmingham to Tamworth road) and is surrounded by roads. There are several gates around the perimeter, any one of which will give reasonable access, and cars can be driven into certain parts of the park itself. For Longmoor Valley enter from Chester Road North (A452 Stonebridge to Brownhills road) at Banners Gate, which is at the junction with Monmouth Drive. Indeed this is perhaps as good as anywhere to begin any exploration of the wilder, more remote parts of the park.

## Calendar

*Resident:* Little Grebe, Great Crested Grebe, Mute Swan, Canada Goose, Mallard, Tufted Duck, Sparrowhawk, Kestrel, Red-legged Partridge, Grey Partridge, Pheasant, Snipe, Woodcock, all three woodpeckers, Skylark, Meadow Pipit, Grey Wagtail and common heath and woodland passerines including Goldcrest, Marsh and Willow Tits, Nuthatch, Treecreeper, Jay, Linnet, Redpoll and Yellowhammer.

*March–June:* Winter visitors leave in March, when Meadow Pipits and Stonechat return. From mid-April, Turtle Dove, Cuckoo, Tree Pipit, Redstart, Whinchat, Grasshopper Warbler, Sedge Warbler, Reed Warbler, Whitethroat, common woodland warblers (including Wood) and Spotted Flycatcher. Passage Wheatears and maybe a Hobby. Possible rarity.

*July–September:* Quiet in July. Departing summer visitors in August and September include Whinchat, Stonechat and Wheatear. Perhaps Oystercatcher, Redshank, Little Gull, Kittiwake, Common Tern, Arctic Tern or Black Tern. Maybe a passing Hobby.

*October–February:* Larks, pipits and Stonechat in October and November, plus possible raptor. Maybe Shelduck (scarce), Wigeon, Teal,

Shoveler, Pochard, Common Scoter (rare), Goldeneye, Goosander or Ruddy Duck. Water Rail and Jack Snipe. Roosting pigeons, thrushes and finches including a few Brambling. Possible rarity.

## Habitat

The Tame Valley lies due south of Tamworth and 9 miles (14 km) to the north-east of Birmingham. It contains a complex chain of pools, marshes, scrub and rough grassland some 10 miles (16 km) long. Although many of the habitats are transient, the ability of birds to move from place to place makes this one of the most important ornithological areas in the whole of the Midlands and arguably one of the most important inland areas in the country.

The valley is an incongruous blend of man-made industrial areas set in pleasant countryside, yet never far from urban influence. Working from south to north the principal sites are listed below.

### SHUSTOKE RESERVOIR

This is a Severn-Trent Water Authority (STWA) reservoir of 100 acres (40 ha) with a steeply shelving, concrete-lined shoreline surrounded by grassy banks. Along the northern shore the River Bourne is lined with alders and conifers. There is some disturbance from sailing and game fishing, but no access to the general public. At times the reservoir acts as an important refuge for wildfowl, especially diving duck. A few Wigeon are regularly present in winter too, and sea-duck are noted quite often.

### LADYWALK RESERVE

This 125 acre (50 ha) reserve is run jointly by the Central Electricity Generating Board (CEGB) and the West Midland Bird Club (WMBC). Situated in a broad meander of the River Tame, the only access is through the Hams Hall Power Station and strictly by permit only. The reserve consists principally of flooded gravel pits which are used to recirculate the cooling water from the power station before it is returned to the river. The warm water inhibits freezing and so the area is very attractive to wildfowl during a hard winter. Some of the old lagoons have been filled with pulverised fly-ash from the power station and these now support a profuse growth of southern marsh orchids each spring. There are also reed-beds, willow scrub and well-vegetated islands to provide plenty of cover for nesting and roosting birds. Little Ringed Plover and a good range of duck breed, while winter brings up to 2,000 wildfowl to the marsh. The recent addition of a scrape promises to make the site even more attractive to waders.

### NETHER WHITACRE

This small reserve is on the opposite bank of the river to Ladywalk and a short distance downstream. It is owned by the Warwickshire Nature Conservation Trust (WARNACT) and, again, it originates from flooded gravel pits, but this time there are dense stands of marginal and emergent vegetation, notably common reed and reedmace, surrounded by scrub. It is thus a favoured site for breeding wildfowl and warblers.

## LEA MARSTON & COTON

Sand and gravel operators are forming two huge river purification lakes for the STWA. The westerly lake, known as Lea Marston, has now been completed and handed over to the STWA. It requires constant dredging and is thus unsuitable for recreational use. Being undisturbed and having a long central island as a safe sanctuary, the lake has become one of the main *focii* for wildfowl in the valley. Indeed, diving duck, notably Pochard, have reached regional records more than once. Extensive tree planting and landscaping have been carried out, but this has still to mature and the site is presently rather bleak and open.

On the other side of the road, the easterly lake, known at Coton, is still under construction. Currently the river divides the excavation, but eventually the river banks will be removed to create one vast sheet of water. The area still resembles a typical gravel pit and, so long as pumping continues to keep the excavations dry, it will remain a favoured haunt of waders as well as duck. Around the gravel workings extensive areas of rough ground act as a well-stocked larder of weed seeds for finches and buntings in winter. The future use of this lake is not known, but meanwhile, it remains one of the prime areas in the valley for birds.

## KINGSBURY WATER PARK

This is the focal point of the Tame Valley wetlands — a fascinating complex of old sand and gravel quarries which has been reclaimed by Warwickshire County Council as a Country Park. In all, Kingsbury Water Park covers over 600 acres (250 ha) and contains more than 20 lakes and pools of various sizes. Recreational pursuits include hydroplane racing, water-skiing, sailing, windsurfing, fishing, camping, caravanning and various casual activities. Yet this is still an outstanding place for birds. The park is dissected by the M42 motorway and the plan is to keep noisy activities and most of the public south of the motorway, reserving the area to the north for more specialised, quieter pursuits. This includes a good nature reserve overlooked by public hides.

The reserve area comprises one main gravel pit, known as Cliff Pool, which has several shingle islands and a small scrape. Water levels can be regulated to some extent for birds, but flooding from the main river cannot be prevented. Adjoining this pool is a restored gravel pit which provides important grazing for Canada Geese and other wildfowl. Immediately south of the nature reserve, both Broomey Croft and Canal Pools hold wildfowl, and two large islands were created in the latter specifically to encourage breeding ducks, waders and terns. In the centre of the park old silt beds and ash lagoons support a vigorous regeneration of willow and alder scrub, with some reedmace and common reed. There are also extensive areas of rough grassland, bramble and hawthorn. Further south are two large and several smaller pools which, though primarily used for active recreation, often hold large numbers of wildfowl. Almost all the birds typical of the valley can be seen in the Water Park, but it is best known for breeding terns, waders and warblers; for wintering wildfowl; and for passage birds, especially in spring.

## MIDDLETON HALL & DOSTHILL

Adjoining Kingsbury Water Park to the north are further large sand and gravel excavations. Currently these are still being worked and parts are kept dry by continuous pumping, making them most attractive to waders. At Middleton there is a smaller, reed-fringed pool and small adjoining wood administered by the Middleton Hall Trust.

## ALVECOTE POOLS

The WARNACT reserve of Alvecote Pools, although situated on the Anker, a tributary of the Tame, is included here because birds often move between this and the above sites. Colliery subsidence has created several shallow pools, giving a total water area of 300 acres (125 ha), through which the river flows. The range of habitats includes extensive reed-beds, fen, marsh, alder and willow carr, scrub and colliery spoil heaps as well as the river and pools, while the surrounding pastures flood in winter. The pools contain several old tree stumps which are much appreciated by passage terns. Generally, this is an open landscape, with treeless, prairie farming to the north, but there are a few scattered oaks, birches, alders and willows. The main Euston to Crewe railway line passes through the site and the M42 motorway skirts it to the east.

## Species

The Tame Valley wetlands will captivate birdwatchers at any time. For variety of breeding birds, they are unsurpassed in the Midlands, and both wintering wildfowl and passage waders are outstanding.

To strangers, perhaps the biggest surprise is to find that typical seaside quartet — Shelduck, Oystercatcher, Ringed Plover and Common Tern — all nesting just about as far from the sea as they could possibly get. Five or six broods of Shelduck appear most years at either Alvecote, Coton, Ladywalk, Lea Marston or Kingsbury. No sooner have they hatched than they all too often disappear, victims of pike or other predators. The two waders fair little better. If not washed out by spring floods, two or three pairs of Ringed Plover normally fledge, but most of the Oystercatcher's breeding attempts seem to be abortive. Indeed, it would be more accurate to describe them as incipient, rather than established, colonists. Not so the Common Tern. They made a couple of sporadic breeding attempts between 1969 and 1974. Then, following the creation of some new islands, three or four pairs bred in 1980 and the next year a dozen pairs raised 37 young. Since then, this has become a well-established colony, with 15 to 20 pairs each year regularly producing 40 to 50 young. Coton, Dosthill and Kingsbury are the best places for Oystercatcher and Ringed Plover, while Canal Pool at

*Common Tern, Ringed Plover and Oystercatcher (from left to right) are among the unexpected birds breeding in the Tame Valley*

Kingsbury is the headquarters of the Common Tern colony. To watch these graceful 'sea swallows' plunging after fish is a delight of summer.

At least six pairs of Little Ringed Plover are spread throughout the valley. An opportunist species, it nests wherever conditions are suitable, but Coton, Dosthill, Ladywalk and, above all, Kingsbury are all regular haunts. Some half-a-dozen pairs of Redshank also nest at Alvecote, Coton, Ladywalk and Kingsbury, either on small islands or in damp pastures. As a breeding species, Snipe is slightly scarcer, but two or three pairs usually nest at Alvecote with others less regularly at Ladywalk and Kingsbury. Several pairs of Lapwing complete the breeding waders of the valley.

Breeding wildfowl are just as outstanding. Small numbers of Little and Great Crested Grebe, Mute Swan, Canada Goose, Mallard and Tufted Duck nest in several places each year. With them, two or three pairs of Pochard raise broods at Alvecote, up to six pairs of Gadwall and Shoveler may be seen, mostly at Kingsbury, and several broods of Ruddy Duck emerge from the shelter of the reeds, particularly at Alvecote and Middleton. Teal nest erratically and sparingly, with Ladywalk and Nether Whitacre favoured areas. These, together with Kingsbury, are also the most likely places to find a Garganey. A drake, or sometimes a pair, of this scarce, elusive duck appears somewhere in the valley practically every spring and autumn. Even summer sightings are by no means rare and breeding has occurred more than once. In fact this is probably one of the most consistently visited areas in the West Midlands.

The valley is also rich in warblers, with nine breeding species. Early in spring every thorn bush and reed-bed seems to reverberate to their songs and this is the ideal place to test your identification skills. Many of these early songsters are merely passing migrants, but there are healthy breeding populations too. Sedge and Willow Warblers are the commonest and most widespread, with the other species more discerning in their choice of habitat. Despite the almost total lack of woodland, Chiffchaffs, Blackcaps and the scarcer Garden Warbler still sing wherever there are a few mature trees with a dense shrub growth beneath them. Scattered thorn thickets make the ideal home for several pairs of Whitethroat, while good numbers of Lesser Whitethroat 'rattle' from deep within the taller, thicker hawthorn hedges. To find Reed Warblers you will need to search out the small patches of common reed, although a few birds do sing from reedmace. Nowhere is this a common bird however. Nor is the Grasshopper Warbler, which has declined markedly in recent times. A decade ago half-a-dozen could be heard from various spots throughout the valley, but now there are seldom more than a couple. Like many migrants, their numbers vary from year-to-year, so it is always worth listening for them in patches of bramble or areas of rank grass. The most productive sites for warblers are Alvecote, Ladywalk, Nether Whitacre and above all Kingsbury, where all nine species can be seen.

A few other breeding species also warrant mention. One or two pairs of Whinchat nest in areas of scrub and rough grassland and at least one pair of Black Redstarts breeds in Hams Hall Power Station. Here and there among the reed-beds Water Rail raise their broods, while along the smaller, unpolluted streams Kingfishers nest, their numbers fluctuating with the severity of the preceding winter. Owls are well represented too, with Little Owls in the older hedgerow trees and Tawny Owls in the small copses and developing woodland. With luck, even a pair of Barn Owls may still survive somewhere in the valley. Certainly the abundance of rough, tussocky grassland must hold enough mammals to provide them with a rich hunting ground. The copses and scrub woodland of the old silt beds hold Turtle Dove, Great and Lesser Spotted Woodpeckers, Spotted Flycatcher, Willow Tit, Jay, Redpoll and Bullfinch. Finally, Cuckoos are widespread, Yellow Wagtails and Reed Buntings nest in the rank grass of damp meadows, and the jangling song of Corn Buntings can be heard across many an open arable field. In cool, wet weather especially, the valley plays host to countless Swifts, Sand Martins, Swallows and House Martins as they skim across the water or hug the sheltered slopes of grassy banks in their search for flying insects.

Once the breeding season is over, return wader passage is eagerly awaited, although it is seldom as good as in spring. Nevertheless, Alvecote, Coton, Ladywalk and Kingsbury can usually be relied upon for small numbers of the commoner species like Oystercatcher, Ringed Plover, Dunlin, Redshank and Common Sandpiper. For Little Ringed Plovers and Green Sandpipers try Kingsbury, where up to 20 often gather in July or August. With them might be a more elegant, speckled Wood Sandpiper. Ruff, Greenshank and Spotted Redshank are also annual visitors to the valley, though not necessarily to every site, while Little Stint, Black-tailed Godwit and Turnstone are fairly regular. Parties of Whimbrel fly through the valley, but are easily missed as they seldom drop down to rest or feed. Curlew are not common either, but a small flock of 20 or so roosts at either Coton or Ladywalk from mid-summer until late autumn and then again in March. Just occasionally a stray Bar-tailed Godwit joins them. Among the less expected autumn visitors of recent years Purple Sandpiper, Pectoral Sandpiper and Grey Phalarope are worth mentioning.

Although few in number, autumn waders frequently stay late, perhaps encouraged by the warm water discharged by the power station. Many species are still present in November and up to 20 Redshank may over-winter in the Ladywalk-Coton-Kingsbury area along with a scattering of Green Sandpipers. Less often a few Ruff join them, usually spending much of the day feeding on fields, but coming into roost on islands at night. Common Sandpipers have also over-wintered along the river more than once, and there are quite exceptional records for the Midlands of Spotted Redshank and Wood Sandpiper having done likewise.

By late summer there are well over 50 Common Terns at Kingsbury and smaller numbers elsewhere in the valley, with perhaps a larger Sandwich Tern or tiny Little Tern as well. Black Terns tend to be spasmodic visitors, either coming in small parties or not at all. Although seldom reaching double figures, they always cause excitement among birdwatchers hoping for a rare White-winged Black Tern — a species which has twice graced Kingsbury in recent years.

The Tame Valley is also the most important area in the region for Mute Swans. At Alvecote, birds begin to gather for their annual moult in June, and by August a herd of over 100 has assembled. Dispersal then occurs during September, when birds begin to congregate at Kingsbury. By October or November almost 100 may be present, mostly on Cliff Pool. Wild swans pass along the valley too, usually parties of Bewick's in late autumn, winter or early spring. Alvecote and Kingsbury are favoured haunts, but quite often birds just drop into a flooded meadow well away from any of the main localities. Whooper Swans have been recorded, but are very rare.

A huge flock of Canada Geese also gathers at Kingsbury in September. It is often joined later in autumn by up to 100 feral Greylag Geese. The vast majority of both species remain in the valley until February or March, during which time they may attract other geese. Mostly these are single Pink-feet, White-fronts or Barnacles of dubious origin, but the occasional small party sometimes suggests genuinely wild birds. Certainly this was the case when, quite remarkably, a party of 20 Bean Geese appeared one March.

Winter wildfowl are impressive both in variety and number. Large flocks of dabbling duck begin to form in August, when up to 600 Mallard are at Ladywalk and 300 at Alvecote. Peak numbers at Kingsbury come later, with typically 200 to 300 in mid-winter. Conversely, Wigeon tend to peak first at Kingsbury, with around 100 in October, after which birds disperse. The highest totals then occur early in the new year, with 150 or so at Ladywalk and up to 50 at both Alvecote and Shustoke. The valley is very attractive to Shoveler, especially in autumn when Kingsbury often holds over 100, Ladywalk 50 and Alvecote 20. One or two Pintail can be seen at various times too. Approaching 50 Gadwall are usually at Kingsbury during the winter, with a late-autumn peak of around 70. Small numbers can also be seen at Middleton.

Recent winters have brought huge flocks of diving duck. Pochard and Tufted now consistently exceed 1,000 and have reached maxima of 2,500 and 1,500 respectively (the former is a regional record). For much of the time these enormous flocks settle at Lea Marston, though the night-feeding Pochard often spend part of the day loafing on Hemling-ford Water. Up to 1,000 Coot are invariably around as well, making this an oustanding site for winter wildfowl. Goldeneye are most numerous at Kingsbury, with between 60 and 70 from January to March. In freezing weather they often take to the river, which is less inclined to ice over. Shustoke, too, holds some 30 Goldeneye during the winter. Small numbers of Ruddy Duck frequent Kingsbury in autumn and one or two Goosanders are often to be seen in winter, either at Alvecote, Kingsbury or Shustoke. Amongst such huge concentrations, keen eyes naturally spot a few rarities. A handful of Scaup are seen most years, but beware as some Pochard hybrids can look very like pure Scaup. Other recent visitors have included Great Northern Diver, Black-necked, Red-necked and Slavonian Grebes, Red-crested Pochard, Ferruginous Duck, Common Scoter, Long-tailed Duck, Smew and Red-breasted Merganser.

Gulls roost in small numbers, mostly at Kingsbury, Ladywalk or, erractically, Shustoke. Many of these spend the day feeding on Packington refuse tip, some 3 miles (5 km) to the south. Most winters one or two Glaucous Gulls are identified, but the rarer Iceland Gull is less regular. Several Water Rail winter among reed-beds, reedmace and other marginal vegetation, notably at Alvecote, Kingsbury and Ladywalk, where very cold weather will bring them out into the open. Snipe also congregate here or in the damp meadows, often in parties up to 100. Jack Snipe are rare or overlooked, but may be forced into the open by heavy snow or extreme cold. If such conditions persist for any length of time even a Bittern may be seen.

The valley is also very attractive to plovers and passerines. Huge flocks of Lapwings up to 8,000 strong are a familiar sight and a well-established flock of about 500 Golden Plover gathers on the arable fields around Drayton Bassett. One or two Rock Pipits may be seen around the lakes in October and November, while in recent years Water Pipits have been increasingly detected. The latter sometimes stay throughout the winter and may be joined by others in March and April. Other passerines congregate around the wealth of disturbed ground, where they feast on weed seeds. Chaffinch, Greenfinch, Linnet, Tree Sparrow, Reed Bunting and Yellowhammer are the commoner species, with perhaps a few Corn Buntings for good measure. Post-breeding and early spring flocks are usually the largest, with the latter sometimes containing a handful of Brambling. Careful searching might be repaid by something really rare, like a Lapland Bunting. Early in autumn, parties of Redwing and Fieldfare strip the hedgerows of berries, then in late winter and early spring they take earthworms from damp pastures. Hedges and fields along the canal are frequently favoured. At Kingsbury and Shustoke, parties of Redpoll and Siskin fidget through the alders, often accompanied by other common woodland birds, while loose groups of Bullfinches are regularly seen at Kingsbury each autumn. A Stonechat may remain somewhere in the area during a mild winter, when a Firecrest or over-wintering Chiffcaff might also be seen.

The wealth of prey attracts raptors. Kestrel and Sparrowhawk are seen most frequently, but often one or two Merlins or even a Hen Harrier will take up residence for the winter. At dusk a Barn Owl, now a rarity in the Midlands, might hunt a patch of rough grassland, while Short-eared Owls search for voles along the rank, tussocky banks separating the many pools. In a really good year as many as twelve of these fascinating owls systematically hunt the area, though just one or two are more usual. Other passing raptors include the occasional Buzzard, usually at Alvecote, or migrant Ospreys in September.

Even in the teeth of a biting cold March wind, small parties of thrushes and Meadow Pipits moving through the valley are a sure sign that spring is on the way. Late in the month the first Oystercatchers, Little Ringed Plovers, Ringed Plovers, Wheatears and Sand Martins may appear. This is a time of great anticipation, for spring invariably brings the best birds. Most of the autumn passage waders such as Dunlin, Ruff, both godwits, Whimbrel, Greenshank, Green Sandpiper, Common Sandpiper and

Turnstone reappear, but in better numbers and reinforced by Sanderling, and perhaps Temminck's Stint, Wood Sandpiper or even Avocet. Recent appearances of Kentish Plover and Broad-billed Sandpiper have further whetted birdwatchers' appetites. With the waders may come a passage of Arctic Terns or perhaps a party of Black Terns. Just occasionally a Sandwich or Little Tern is seen in spring as well.

Early May is usually best for waders and terns, but late May or June are when a rare, exotic heron might overshoot and come drifting in on a warm, southerly breeze. Spoonbill, Night Heron and Little Egret have all done so in the past. One or two Marsh Harriers are now expected each spring, but they seldom stay for long. Neither do the migrant Ospreys that are occasionally seen heading northwards. Notable among the large influxes of passerines each spring are the movements of Pied Wagtails in April, which invariably bring one or two Whites with them, and the Yellow Wagtails that follow, with just a chance of a Blue-headed among them.

## Timing

The beauty of this chain of wetlands is that birds disturbed from one site simply move elsewhere. Timing is not therefore especially critical. For those who like peace and quiet though, mornings are best. Sailing on Shustoke often empties the reservoir of birds. At times Kingsbury gets extremely busy, even in the winter. Traffic noise from the M42 and from hydroplane racing can also be a nuisance to those trying to listen to birds. Fine weekends and Bank Holidays are the busiest times, particularly in summer.

In spring, cold, stormy weather often grounds overflying waders, terns and passerines, while a warm, southerly airstream may bring an overshooting 'southern' heron or other rarity. Swifts and hirundines are most numerous on cool, damp summer days. Often winter wildfowl are more plentiful in very cold weather, when they are also forced to concentrate into a few unfrozen spots. Bittern, Water Rail and Jack Snipe are more evident in such conditions as well. For late waders and passerines a mild winter is generally best. The whole valley complex is prone to flash flooding after heavy rain at any time of year and this can prevent access to some parts.

## Access

If approaching on the M42 Birmingham to Nottingham motorway leave at the A5 junction for Alvecote, or the A446 junction for all other sites.

For Shustoke Reservoir, take the A446 southwards and then turn eastwards at the first roundabout onto the B4114 Birmingham to Nuneaton road. Continue through Coleshill and on to Shustoke, about 2 miles (3 km). Turn left in the village to the reservoir. Access is strictly by permit only from the STWA.

For Ladywalk, take the A446 southwards and then turn left in about ½ mile (0.8 km) towards Lea Marston. Where the road turns sharp left, continue straight ahead, over the railway line and into the main entrance of Hams Hall Power Station. Again entrance is strictly by permit only, available from the WMBC to members only. Within the reserve are hides overlooking the marsh.

For Nether Whitacre, take the same route to Hams Hall, but follow the road round to the left at the entrance to the power station and continue into Lea Marston village. Turn right in the village, cross over the river, pass the entrance to the purification lake on your left (it may be worth a quick check of the lake from here) and over the railway line. The reserve is then on the right-hand side. Access is restricted to WARNACT members only and is through a locked gate.

For Lea Marston and Coton, take the A4097 towards Kingsbury. Turn right at the first roundabout and continue for about ½ mile (0.8 km). Park on the verge, but take care as this road is busier than it looks. Lea Marston purification lake is then on the right-hand side and can be viewed from the road. Mornings are best for this, as later in the day you will be looking into the light. There is a hide here, but arrangements for its use have still to be finalised. The lake itself is an important operational area to which there is strictly no access. Coton is on the opposite side of the road. At present this is still being worked for sand and gravel and it, too, is strictly private. The workings are divided into two parts by the river. The northern part can be viewed quite adequately from the road. For the southern part, which is best for waders, there is a public footpath which leaves the road on the left immediately before the railway bridge. Follow this, cross over the stream and much of the area can then be viewed from here. Because of the distances involved, a telescope is desirable at both sites.

For Kingsbury Water Park, take the A4097 towards Kingsbury, but turn left at the first roundabout. Follow the lane round to the left and the main entrance to the Water Park is on the right in about ½ mile (0.8 km). For the nature reserve area continue past the main entrance and over the motorway. Then turn immediately right into a tiny lane and take the entrance on the left signed to Broomey Croft. Follow this road, forking right to leave the caravan site on your left, pass through the lifting barriers and continue to the car park at the end. Then follow the paths northwards to the two hides. There is a whole system of paths throughout the park, most of which are worth exploring, especially in summer. A charge is made for parking at both ends of the park, so it is generally better to go into Broomey Croft car park and explore from there. For those using the main entrance, Far Leys car park is probably the most convenient.

For Dosthill, park in the Broomey Croft car park at Kingsbury and walk northwards along the canal towpath. The gravel pits are on both sides after about 1 mile (1.6 km). Again there is no direct access, but areas can be seen from the canal, or from the accommodation bridge across it. Further north, at Fishers Mill, another bridge carries a footpath through the area. This footpath leaves the A4091 Birmingham to Tamworth road to the right just before the end of the dual carriageway when travelling north.

For Middleton, take the A446 northwards from the M42 junction and in 1 mile (1.6 km) turn right onto the A4091 towards Tamworth. In a little over 2 miles (3.5 km), opposite a turn to Middleton Village, turn right into Middleton Hall. The Hall and the nature trail around the lake are only open to visitors on Sunday afternoons. At other times view from the road.

For Alvecote, take the A5 westwards from the M42 junction and then turn right towards Polesworth. Follow the road through a new housing estate until you reach the B5000. Turn right onto this road and then left towards Alvecote. In about 1 mile (1.5 km) there is a small picnic area on the right immediately before the canal bridge. Park here, cross over the canal and follow the waymarked trail to the right to explore the

Pooley Field part of the reserve to the south of the railway line. For the northern area, cross over the canal and railway line, pass through Alvecote village and turn either right or left at the end. Good views can then be had from this road, which passes between the larger pools. For access to the reserve itself, contact WARNACT.

## Calendar

*Resident:* Little Grebe, Great Crested Grebe, Mute Swan, Greylag Goose (feral), Canada Goose, Shelduck, Gadwall, Teal, Mallard, Shoveler, Pochard, Tufted Duck, Ruddy Duck, Sparrowhawk, Kestrel, Water Rail, Coot, Lapwing, Snipe, Redshank, Barn Owl (rare), Little Owl, Tawny Owl, Kingfisher, Great Spotted Woodpecker, Lesser Spotted Woodpecker and common passerines including Redpoll, Bullfinch, Reed Bunting and Corn Bunting.

*March–June:* Winter visitors depart in March and April. Garganey (scarce), Oystercatcher, Little Ringed Plover, Ringed Plover, Sanderling, Dunlin, Ruff, Black- and Bar-tailed Godwits, Whimbrel, Curlew (small roost in March), Greenshank, Green Sandpiper, Wood Sandpiper, Common Sandpiper, Turnstone, Sand Martin, Meadow Pipit (March), Water Pipit (rare), White Wagtail and Wheatear. From April onwards, Common, Arctic and maybe Sandwich, Little and Black Terns, Turtle Dove, Cuckoo, Swift, hirundines, Yellow Wagtail, Black Redstart, Whinchat, Grasshopper Warbler, Sedge and Reed Warblers, scrub and leaf warblers and Spotted Flycatcher. Maybe passing raptors. Possible rarity.

*July–September:* Summer visitors depart. Flocks of Lapwing and common wildfowl, especially moulting Mute Swans. Garganey (scarce), Oystercatcher, Little Ringed Plover, Ringed Plover, Little Stint, Dunlin, Ruff, Black- and Bar-tailed Godwits, Whimbrel, Curlew (small roost), Spotted Redshank, Greenshank, Green Sandpiper, Wood Sandpiper, Common Sandpiper, Turnstone and terns perhaps including Sandwich, Little or Black. Maybe passing raptors. Possible rarity.

*October–February:* Bewick's Swan, Wigeon, Pintail (scarce), Scaup, Goldeneye, Goosander, Hen Harrier (scarce), Merlin, Golden Plover, Ruff (scarce), Jack Snipe (scarce), Curlew (autumn roost), Green Sandpiper, roosting gulls including, after December, Iceland (rare) or Glaucous (scarce), Short-eared Owl, Rock Pipit (October and November), Water Pipit (scarce), Stonechat, Fieldfare, Redwing, Siskin and mixed flocks of Tree Sparrows, finches and buntings. Possible diver, rare grebe, Whooper Swan, party of grey geese, sea-duck or Smew. Maybe a passing raptor, late wader or rarity.

# BENTLEY PARK, MONKS PARK & HARTSHILL HAYES

Map 19
OS map 140

## Habitat

These relics of primary oakwood are reputed to be remnants of the old Forest of Arden. Each has been much altered over the years, but at least Bentley Park and Monks Park have kept their broad-leaved character. Here, on a freely draining plateau, acidic soils support oak-birch woodland with an understorey of bracken and bramble. In the wetter areas there are also ash, alder, hazel and elder, while both beech and sycamore are locally dominant.

Much of Hartshill Hayes on the other hand has been clear-felled and replanted by the Forestry Commission (FC) with larch, pine and spruce. Useful pockets of broad-leaved timber have survived, however, including several large oak standards that formerly rose above extensive coppice. An interesting feature of this wood are the limes that used to be coppiced for the hat-makers of nearby Atherstone. Adjoining the Hayes is open, rough grassland and a small, mixed wood of pines and sycamore.

## Species

The breeding community includes a good range of woodland species, without having any star attractions.

Commoner residents like Wren, Dunnock, Robin, Blackbird, Song Thrush, Blue and Great Tits, and Chaffinch are widespread and numerous. They ought to be seen on practically any visit. Other residents are more scarce or selective in their choice of habitat, and so take more finding. Treecreepers are thinly spread, but unobtrusive. In winter especially, they often attach themselves to wandering parties of tits. Nuthatches, too, are thinly spread amongst the older oaks and beeches. In springtime they can easily be located by their ringing calls, which echo through the trees. This is also the best time to find woodpeckers. There are a few Green Woodpeckers around and their laughing calls can be heard at any time. A favourite spot is the rough pastures adjoining Hartshill Hayes. But the other two woodpeckers are most easily detected when they are drumming. Great Spotted is easily the commoner of the two and drums the louder. In spring it can often be found hammering away at rotting timber, frequently just beneath a bracket fungus. The Lesser Spotted is the scarcest of all the woodpeckers and the hardest to locate as it flits through the crowns of large oaks, often making its nest-hole in a branch rather than the trunk.

Before long the resident birds are joined by summer migrants. Chiffchaffs arrive first, closely followed by Willow Warblers. The lofty, standard oaks of Hartshill Hayes seem especially to attract Chiffchaffs, whereas the cheerful song of Willow Warblers can be heard almost everywhere. Blackcaps occur in reasonable numbers, favouring the valleys where there is a good, thick shrub layer beneath the canopy, but Garden Warblers are fewer and generally confined to thick scrub in open glades and clearings. The more open parts and woodland edges hold a few Whitethroats and Spotted Flycatchers. Less fastidious are Jays and

Turtle Doves, whose raucous screams and soft purring emanate from broad-leaved and coniferous woodland alike. The conifer plantations sustain Goldcrests and Coal Tits in plenty, along with a few pairs of Redpoll. In winter they are also used by flocks of roosting thrushes and finches.

Bentley Park and Monks Park generally have the wider range of migrants. Here one or two Wood Warblers sing from within the oakwood and the floating song flights of Tree Pipits can be watched around clearings and rides. With luck, even a pair of Redstarts may be glimpsed in courtship display. Once a passage Pied Flycatcher stayed for a few days.

Occasionally Woodcock are flushed from the bracken in broad daylight, but they are much more likely to be seen at dusk, either in roding flight or leaving their roosts to feed. One or two pairs breed, but many more are present in winter. Overhead, Kestrel and Sparrowhawk are the usual daytime raptors, but several Little Owls emerge to hunt at twilight and, as darkness deepens, the hoots of Tawny Owls echo eerily in the chill night air.

## Timing

As with any woodland, early on a bright spring morning is the best time to visit. Song is then at its height, making the birds easier to find. April, before the leaves have opened, offers the best chance to see resident birds, but for summer migrants May is best. By high summer, most species are difficult to locate. In winter birds are active throughout the short days, frequently roaming together in loose flocks. Stumbling across such a flock is very rewarding, but failing to do so can be very frustrating. Hartshill Hayes gets busy in the afternoons, and is best avoided on fine weekends and school holidays.

## Access

Bentley Park and Monks Park are private woodlands, but fortunately some of the better areas can be seen from public rights-of-way. On no account should birdwatchers leave these and trespass into the woods. Approach from the B4116 Atherstone to Coleshill road. Park on the roadside verge at Bentley and there is a footpath into the wood opposite the public house. This can be followed through to either Ridge Lane or Birchley Heath, or a circular walk made by using the road between the two ends of the footpaths.

Hartshill Hayes (now a Warwickshire County Council Country Park) is best approached from the B4114 Birmingham to Nuneaton road, turning northwards in Hartshill and then left opposite the school playing field. After ½ mile (0.8 km) turn right into the Country Park where there is a car park. Follow the waymarked paths through the wood.

## Calendar

*Resident:* Sparrowhawk, Kestrel, Woodcock, Little Owl, Tawny Owl, all three woodpeckers and common woodland birds including Goldcrest, Nuthatch, Treecreeper, Jay and Redpoll.

*April–June:* Remaining winter visitors leave in April. Turtle Dove, Tree Pipit, Redstart, Whitethroat, woodland warblers including Wood Warbler, and Spotted Flycatcher. Perhaps a scarce migrant.

*July–September:* July is quiet as family parties stay well hidden. Summer visitors leave during August and September.

*October–March:* Roosts of thrushes (including Fieldfare and Redwing) and finches (including Brambling).

# DRAYCOTE WATER

Map 20
OS maps 140 & 151

## Habitat

Draycote Water is a water-supply reservoir administered by the Severn-Trent Water Authority (STWA) and lying 4 miles (6.5 km) to the south-west of Rugby. Covering 700 acres (280 ha), this is the second largest water in the region. Much of its shoreline comprises earth dams, which are grassed to landward, surmounted by a perimeter road and stone pitched against the water to prevent erosion. The most extensive natural shoreline is the northern one, which shelves quite steeply. However, there are two or three bays along it which are shallow and sheltered enough for marginal vegetation to establish. Of these, Toft Bay in the extreme north-eastern corner is best for birds.

Recreational activity is intense, with the largest inland sailing club in Britain, a flourishing game fishery from March to October, windsurfing, canoeing and occasionally sub-aqua diving and sponsored walks. These, plus the scarcity of marginal or emergent vegetation, and the absence of islands or a designated sanctuary, severely limit breeding species. In all other respects, though, this is one of the best places in the region for birds.

Draycote is replenished by water pumped during winter from the nearby River Leam. Consequently the water level is highest in spring, so wader passage is meagre, and lowest in autumn, when more waders may appear. However, it is from October to March that the birdlife is at its best. Then there are outstanding concentrations of wildfowl and gulls. Like other large waters in the region, Draycote also has a penchant for turning up rarities.

Many trees have been planted around the reservoir, and as these are maturing so the variety of passerines is increasing. The surrounding grassland, which is now grazed by sheep, attracts feeding wildfowl and passerines in winter. Outside the reservoir confines, but visible from it, is a small sewage works. Effluent here is discharged through pipes onto a sloping field, which in consequence is very marshy with a good stand of reedmace. This marsh is good for birds, especially in winter.

## Species

Winter wildfowl are the most interesting and varied group. Numbers begin to build up from September onwards and are normally good throughout the winter, declining again the following March. Often they reach their maximum in severe weather, when many smaller waters are frozen solid.

Draycote is strong in some species, but weak in others. Overall it is a water for diving rather than dabbling duck. Yet there are some important exceptions even to this. Mallard typically reach 800 in mid-winter and between 250 and 400 Wigeon are regularly present, attracted no doubt to the grazing provided by the grass surrounds. On the other hand, the lack of shallows and emergent vegetation for cover keeps other species down. For example, Mute Swans and Canada Geese are hardly ever recorded. Teal nevertheless regularly reach 150 to 200 around the turn of the year and good numbers of Gadwall are quite common in late autumn, with 25 a normal count and over 100 on record. Small parties of three or four Shelduck pass through each spring and autumn and up to a dozen Shoveler appear intermittently during the winter. One or two parties of Bewick's Swans or a skein of White-fronted Geese pass through most years with December a likely time for both to appear, and swans often returning in March. From Hensborough Bank, check the river meadows to the south as Wigeon, geese and swans sometimes settle there to graze. Whooper Swans are rare visitors.

Diving duck are much more numerous. Peak counts often occur in mid-winter, when there are typically 150 Great Crested Grebes, 300 Pochard and Tufted Duck, 150 Coot and 60 Goldeneye. At times much larger numbers congregate, and Pochard, Tufted Duck and Coot have all been well in excess of 1,000 in the past. Considering these numbers, Goosander are surprisingly scarce, with seldom more than two or three which rarely stay right through the winter.

Draycote has a reputation for scarce and rare wildfowl. Smew is a regular visitor from December through to March. Most are females or immatures, but a snow white drake occasionally occurs. Red-breasted Merganser is nearly as common as Goosander. Indeed, during a recent freeze there was the delightful spectacle of both sexes of all three sawbills together. Sea-duck, too, are regular winter visitors, with some individuals making protracted stays. Recently Scaup, Eider, Velvet Scoter and Long-tailed Duck have all stayed for several months. Conversely, Common Scoter, though occurring more regularly, seldom stay as long. Sawbills and sea-duck often prefer the western end of the reservoir.

Divers are another of Draycote's specialities. Great Northern Diver, especially, is an almost annual visitor, with one or two usually arriving in November or December and sometimes staying a few weeks. Black-throated Divers are more erratic, often arriving during spells of hard weather on the Continent. Red-throated is the scarcest of the three.

*Divers, such as this Black-throated Diver, are one of Draycote's specialities*

The gull roost is huge, with estimates of over 100,000. Some 80 per cent or more are Black-headed Gulls, which makes the search for the similar, but rare, Mediterranean Gull extremely difficult. Nevertheless one or two are very infrequently detected. Common Gulls are also more numerous here than at many Midland waters. Among the larger gulls,

Lesser Black-backed and Herring are quite plentiful and over 100 Great Black-backed Gulls are by no means unusual. Although scarcer than at some waters, one or two Glaucous and Iceland Gulls are increasingly appearing, especially on colder nights.

In March wildfowl depart and fishing commences. Anglers and a high water level normally portend a poor wader passage, though there are notable exceptions. In either case birds pass through very rapidly. Oystercatcher, Little Ringed Plover, Ringed Plover, Dunlin and Common Sandpiper are the more regular waders, though in numbers less than ten. Less regular are Grey Plover, Sanderling, Knot, Whimbrel, Greenshank and Turnstone. Occasionally there is a real surprise, like a party of over 50 Bar-tailed Godwits or a splendid Spotted Sandpiper in full breeding plumage.

Less affected by anglers, terns and gulls are often just as interesting as waders in spring. A few Little Gulls and Kittiwakes pass through most years, while late April and early May bring the main passage of terns. Arctic Terns are the most numerous, with parties of 50 or more sometimes appearing in deep depressions. By comparison, Common Terns are few. A small flock of Black Terns might also pass through, though they are very erratic. Scarcer, but more reliable, are Sandwich and Little Terns, with one or two most years.

Of the migrant passerines, April brings a trickle of Wheatears and a few White Wagtails, followed by good numbers of Yellow Wagtails. The grassy banks and the sewage works filter beds are good places for wagtails. By May, Cuckoos are well established, Sedge Warblers sing from the willow scrub, a pair of Reed Warblers usually takes up residence in Toft Bay, and Blackcap, Whitethroat and Lesser Whitethroat can all be heard singing from scrub or thick hedges. A Grasshopper Warbler might also be heard from within a patch of bramble or tussocky grass.

By high summer the reservoir is deserted, save for a few broods of Mallard and Tufted Duck, and hordes of Swifts and hirundines feeding on the myriad of midges. Soon young Grey Herons begin to arrive from the nearby heronry to fish. In July they are joined by the first Common Sandpipers, with perhaps a solitary Green Sandpiper or an early Ringed Plover or Dunlin. August sees a steady trickle of waders. Ringed and Little Ringed Plovers, Dunlin and Common Sandpiper are the commonest, but even they seldom exceed a dozen, although from time-to-time twice as many are noted. This is a favourite staging post for Greenshank, though, with up to ten regularly present during August and September. Three or four Ruff and a couple of Oystercatcher, Sanderling and Turnstone are also seen each year, while one or two juvenile Little Stints usually turn up in September or October. More rarely, an adult might be seen in late July or early August. Curlew Sandpipers are much scarcer and more erratic. Occasionally a party of waders arrives, for example flocks of 21 Curlew and 22 Bar-tailed Godwits. As a rule though, these species along with Knot, Black-tailed Godwit, Whimbrel, Spotted Redshank and Wood Sandpiper are scarce and irregular. Late autumn may bring a few more Dunlin, some of which could stay well into winter. With them may be one or two Ruff and a solitary Green Sandpiper or Grey Plover, while recent rarities have included Grey Phalarope, Spotted Sandpiper in autumn dress, and Long-billed Dowitcher.

Autumn is not just a time for waders. Terns pass through once more on their way south, but this time Common Terns outnumber Arctic. Again a Sandwich or Little Tern, or a party of Black Terns is possible. Terns are sometimes accompanied by a few Little Gulls or Kittiwakes, and with Black Terns there is always the hope of White-winged Black Tern — a rarity which has made several appearances here. This is also the time to hope for a migrant Osprey coming to fish.

Of the passerines, Wheatears and Yellow Wagtails pass steadily through and by October the first Redwing and Fieldfare arrive just as the last Swallows and House Martins are leaving. A few Siskin may settle into the alders along the northern shore. Late autumn usually produces something of interest, be it a Water Rail, Jack Snipe or small party of Bearded Tits in the sewage works, a couple of Snow Buntings along one of the dams, or a Rock Pipit working its way around the shoreline.

Each year autumn gales are eagerly awaited by birdwatchers. Strong westerlies in September or October almost invariably bring a seabird or two. Shags are commonest, but other recent vagrants have been Fulmar, Manx Shearwater, Leach's and Storm Petrels, and Gannet. Skuas and gulls may also be driven inland, and Arctic, Great and Pomarine Skuas have all appeared over the years. Pride of place, though, must go to a beautiful Sabine's Gull that graced the reservoir a few years back.

In winter passerines may resort to the shoreline or the rough, grassy banks, though numbers recently have been low. Tree Sparrows, Chaffinches, Greenfinches, Goldfinches, Linnets and Yellowhammers are most regular, but exceptionally they are joined by a few Brambling or even a Twite or two. A flock of Stock Doves is often present, Green Woodpeckers are commonly seen and a Grey Wagtail might frequent the sewage works or the small stream beneath Draycote dam. A few Reed Buntings and Meadow Pipits are usually present as well. There are few overhanging perches for Kingfishers, but they do appear when other waters are frozen over.

With the abundance of prey, raptors are frequently seen. Sparrow-hawk and Kestrel are commonest and both breed nearby, while Hobbies are frequently seen in spring and late summer. In autumn or winter a Merlin or Short-eared Owl might stay around for a while to further enliven the birdwatching at this excellent site.

## Timing

So far Draycote has not been opened up to general public access, so the main disturbance comes from sailing, windsurfing and angling. The worst times are when sailing and fishing are both taking place. The fishing season runs from late March through to October, and takes place both from bank and boat. Bankside activity is greatest during the first few weeks, but is not too great during the autumn. Fishing from boats seems to be more or less uniform throughout the season. Sailing takes place mostly during weekend afternoons.

Morning visits are generally the best, except for roosting gulls when the very last hour of daylight is better as many birds come very late into the roost. Extremely cold winter days often yield the highest wildfowl counts and cold nights usually bring the most gulls. In spring, many migrants arrive during overcast conditions with light southerly winds, but Arctic Terns and waders are quite frequently brought in by strong winds and deep depressions. In autumn, try a visit a day or two after a vigorous depression has crossed the Atlantic. The wake of hurricanes are always a good time.

## Access

The only entrance to Draycote Water is on the right-hand side of the A426 Rugby to Southam road, almost 2 miles (3 km) south of Dunchurch. There is strictly no access to the reservoir except for permit holders.

Permits, available on an annual basis for a reasonable charge, can be obtained from the garage in nearby Kites Hardwick. This is on the right-hand side about ½ mile (0.8 km) further on towards Southam. The permits entitle holders to drive around the perimeter road and to use their cars as mobile hides. Even so, the sheer size of the reservoir makes a telescope essential for complete coverage.

For casual access there is a small country park on Hensborough Hill, which overlooks the reservoir from a distance. Access is the same as for the reservoir and there is a car park for which a charge is made. From here a small section of shoreline can be reached on foot. There is also a public footpath which skirts the north-eastern perimeter of the reservoir. However, neither the views from this footpath, the short length of public foreshore, nor the country park are adequate for birdwatching, even with a telescope.

## Calendar

*Resident:* Grey Heron, Mallard, Tufted Duck, Sparrowhawk, Kestrel, all three woodpeckers, Pied Wagtail and most other common passerines.

*March–June:* Winter visitors depart in March. Shelduck (scarce), Common Scoter (scarce), Oystercatcher, Little Ringed Plover, Ringed Plover, Grey Plover (scarce), Knot (scarce), Sanderling (scarce), Dunlin, Bar-tailed Godwit (scarce), Whimbrel (scarce), Redshank, Greenshank (scarce), Common Sandpiper, Turnstone (scarce), Little Gull and Kittiwake (scarce), Sand Martin, Meadow Pipit and Wheatear. From mid-April Hobby (scarce), Common Tern, Arctic Tern, perhaps Sandwich, Little or Black Tern, Cuckoo, Swift, hirundines, Yellow and White Wagtails (sometimes Blue-headed), Whinchat and warblers including Grasshopper, Sedge, Reed and Lesser Whitethroat. Possible rarity.

*July–September:* Departing summer visitors include Swift (mostly July and August), hirundines, Yellow Wagtail, Whinchat, Wheatear and warblers. Shelduck (scarce), Gadwall, Common Scoter (scarce), Hobby (scarce), Osprey (rare), Oystercatcher, Little Ringed Plover, Ringed Plover, Knot (scarce), Sanderling (scarce), Little Stint, Curlew Sandpiper (rare), Dunlin, Ruff, Snipe, Black- and Bar-tailed Godwits (scarce), Whimbrel (scarce), Curlew (scarce), Spotted Redshank, Redshank, Greenshank, Green Sandpiper, Wood Sandpiper (rare), Common Sandpiper, Turnstone, Little Gull and Kittiwake (scarce), terns including perhaps Sandwich, Little or Black. Probable vagrant seabird in September. Maybe other rarity.

*October–February:* Little and Great Crested Grebes, Mute Swan (scarce), Bewick's Swan (scarce), Whooper Swan (rare), White-fronted Goose (scarce), Wigeon, Gadwall, Teal, Pintail (rare), Shoveler (scarce), Pochard, Goldeneye, Smew, Red-breasted Merganser (scarce), Goosander, Merlin (scarce), Coot, Water Rail, Golden Plover, Grey Plover (October, scarce), Lapwing, Dunlin, Ruff (scarce), Jack Snipe (scarce), Snipe, Green Sandpiper (scarce), roosting gulls including maybe

Mediterranean (rare), or after December Iceland (rare) or Glaucous (scarce), Short-eared Owl (scarce), Rock Pipit (October: scarce), Grey Wagtail, Fieldfare, Redwing, sometimes Siskin, and flocks of sparrows, finches and buntings, with perhaps Snow Bunting (November). Probable diver or sea-duck. Possible rarity.

# BRANDON MARSH

Map 21
OS map 140

## Habitat

Brandon Marsh covers an area of 132 acres (54 ha) of old colliery subsidence pools and gravel pits on the outskirts of Coventry. It is owned by Steetly Construction Materials Limited. The habitat ranges from open water, through reed-beds and willow scrub, to two small areas of woodland. With gravel extraction still active, some habitats are transient, but much of the area has been very actively and successfully managed by local conservationists for 20 years. Given the security of a recent agreement between the owners and the Warwickshire Nature Conservation Trust (WARNACT), some ambitious projects are being undertaken and the area is constantly being improved.

The original colliery flash lies in a meander of the River Avon, to which it is connected at its western end. As a result, the water level fluctuates markedly. In winter it is often so high that the surrounding marsh and sallow swamp, and the footpaths through them, are under several inches of water. In summer it may be so low as to leave a large expanse of mud with little open water. Such fluctuations greatly increase the variety of birdlife. Attempts are therefore being made to regulate the water levels in the adjacent Teal Pool and Eastern Marsh Pool for the benefit of birds. At the same time, two substantial islands have been created in the Eastern Marsh Pool to provide wildfowl with safe breeding, loafing and roosting sites. Some of the flooded gravel pits also have several small, well-vegetated islands which act as a sanctuary for wildfowl, while others are used as silt settling beds. These quickly become colonised with a dense growth of sallow and osier that makes an ideal roost site.

There are also areas of rush, sedge, reed-grass and reedmace, plus a good stand of common reed that over the years has consistently attracted interesting birds. On the edge of the reserve are two small coverts, one of mixed woodland and one of conifers. Between them, these house a few woodland species that add further to the diversity of this interesting site.

## Species

Wildfowl, passage waders, warblers and Bearded Tits in winter are the main attractions.

Among the wildfowl, Teal are especially numerous. There are over 200 from September through to November or December, by which time a peak of 400 is quite likely and as many as 700 have been recorded. After the New Year, numbers depend on the severity of the weather. In a mild winter a couple of hundred may stay until February or early March, but if the marsh is frozen solid few will be seen until the return passage in March. Mallard numbers, by comparison, are small, with a peak of 200 to 300 in late summer, but often less than 50 through the winter. A few Wigeon are normally present throughout the winter, with occasional influxes resulting in 100 or so for a few days. Small numbers of Shoveler and Gadwall can generally be seen throughout the year, with the former reaching 75 or so in late summer. Gadwall may even attempt to breed, as might Shelduck, two or three of which appear almost every spring. One or two Garganey are also seen most years, both in spring and late summer, and breeding has occurred in the past.

Of the diving duck, Tufted is generally the commonest, with numbers steadily increasing through the autumn to a winter peak of just over 100. A few pairs breed and their young, together with a small influx from elsewhere, often create a subsidiary peak in July. Small numbers of Pochard are always present, except in high summer, but the main peak of 50 or so is usually reached in mid-winter. A few Goldeneye are also likely to winter, but sawbills are irregular, although all three species have occurred, with Goosander most frequent. One or two pairs of Ruddy Duck breed and a solitary individual may over-winter if conditions are not too harsh.

Two or three pairs of Little and Great Crested Grebe also breed and small numbers of both species are usually present in winter, while a Black-necked Grebe is occasionally seen in spring. Canada Geese are invariably around in good numbers, with some 15 breeding pairs, up to 200 in late summer and 100 or thereabouts through the winter. Mute Swans, though, have little breeding success and seldom number more than a dozen.

As the winter wildfowl depart, wader passage begins. The first Ringed Plovers, Dunlin, Little Ringed Plovers and Redshank are often seen in March, with one or two pairs of the latter two species settling in to breed. By April and May a steady stream of waders will be moving through. Numbers are small, but variety can be good with perhaps a few Ruff, Curlew, Spotted Redshank in their dark breeding dress, Greenshank and Green and Common Sandpipers to add to the above species. In May, one or two Sanderling might also be seen.

Post-breeding flocks of Lapwing begin to form again in late June and by July up to 1,000 may be present on the marsh. July also sees the start of the return wader passage, when Green Sandpipers reach a peak of a dozen or so and the first Common Sandpipers appear. August sees most movement, with Ringed Plover, Dunlin, Ruff, Spotted Redshank and Greenshank almost guaranteed, and Whimbrel, Wood Sandpiper or an early Grey Plover possible. To these may be added a Little Stint or even a Curlew Sandpiper in September. It will also see an influx of Snipe, up to 30 of which will then remain until the following April. One or two Jack Snipe usually pass through in October or November, then return again the following March. Just occasionally, when the weather is mild, one remains throughout. A solitary Dunlin, Green or Common Sandpiper might also linger well into November.

Terns are scarce, but most of the common species pass through particularly in late spring. Of the gulls, only Black-headed is numerous, with around 1,000 in autumn and a distinct possibility that one day breeding will be established.

Kingfishers are frequent visitors at all seasons. On the East Marsh, Lapwing, Redshank and Snipe breed in the damp, marshy hollows and Cuckoos keep a careful watch on their host species. This is an excellent spot for raptors too. Apart from the resident Kestrels, a Marsh Harrier passes through almost every year in April or May. Later in the year, a Hen Harrier is often seen in November or December, sometimes quartering the same area as a Short-eared Owl, while with luck a roosting Long-eared Owl might emerge to hunt at dusk.

The reed-beds hold up to 50 pairs of Reed Warbler and perhaps as many as six breeding pairs of Water Rail. The latter, though, are much more evident in winter, especially during hard weather. The elusive Spotted Crake has also been recorded here more than anywhere else in the region, though not in recent times. Most occurrences have been between August and November. This is also the only place in the West Midlands where Bearded Tits are at all regular, with a few arriving most years in October and November and sometimes staying through until March. These arrivals coincide with autumnal irruptions from the East Anglian and Continental breeding grounds.

The sallows are home to several pairs of Sedge Warbler and to some 250 roosting Pied Wagtails in autumn. They used to hold thousands of Swallows too, but recent roosts have contained only a few hundred birds. The numbers of Reed and Corn Bunting roosting on the marsh have likewise fallen in recent winters. With the reduction in Swallows, Hobbies are making fewer visits, although they are still reasonably regular throughout the summer. The two coverts hold all three woodpeckers, Tawny Owl and probably one or two Woodcock in winter.

In spring the whole area is alive with bird song as Turtle Dove, Blackcap, Garden Warbler, Whitethroat, Lesser Whitethroat and Spotted Flycatcher return to defend their territories in wood or scrub. From somewhere on the marsh the mechanical reel of a Grasshopper Warbler can usually be heard as well. In winter the marsh becomes a feeding and roosting ground for small flocks of thrushes, finches and buntings. Amongst them a few Redpoll and Siskin can usually be found in the alders of the Central Marsh, pursued perhaps by a Sparrowhawk twisting and turning through the trees.

Most of the common migrants like Yellow Wagtail or Wheatear pass through each spring and autumn and sometimes a scarcer one like Redstart, Ring Ouzel or Pied Flycatcher. Just occasionally something even rarer turns up, often in a ringer's mist-net! In the past both Great Reed Warbler and Barred Warbler have been discovered this way. Not all of the best birds end up in nets, though, and recent sightings of a winter Bittern and an autumn Little Bittern both serve to show the potential of this excellent marsh.

## Timing

Timing is not too important from the disturbance point of view. Nevertheless, early mornings are still the best time, especially for songsters. In summer raptors are more active towards mid-day, while late afternoons can be good in winter for raptors and owls as well. Most of the open water freezes solid in hard weather, forcing many birds to forsake the area. A southerly wind in late spring may well bring a rarity.

## Access

The site is managed by WARNACT and access is restricted to members only. The marsh lies on the south side of Brandon Lane, 3 miles (5 km) south-east of Coventry, and is best approached from the A45 North-hampton to Coventry (Birmingham) road. Entry to Brandon Lane is from the eastbound carriageway only, so if you are approaching in a westerly direction, go right around the roundabout at the A423 junction into Coventry and retrace your route back towards Northampton. Brandon Lane is almost immediately on your left. Follow the lane for just over 1 mile (2 km) and then park on the verge by the main entrance to the gravel pit on the right-hand side. On no account take cars into the quarry even if the gates are open, or you might later find yourself locked in.

Follow the track down past the quarry buildings and on between two pools until you come to a Sand and Gravel Association commemorative plaque and a WARNACT notice board. From here leave the main track as it bears left and strike out straight ahead along a well-defined footpath. This leads to viewing screens and the two main hides. For other paths around the area see the map on the WARNACT notice board.

## Calendar

*Resident:* Little and Great Crested Grebes, Mute Swan, Canada Goose, Gadwall, Mallard, Tufted Duck, Ruddy Duck, Sparrowhawk, Kestrel, Water Rail, Coot, Lapwing, Snipe, Tawny Owl, Kingfisher and common passerines including Jay, Redpoll and Reed Bunting.

*March–June:* Winter visitors depart in March. Black-necked Grebe (rare), Shelduck, Teal (especially March), Garganey (scarce), Marsh Harrier (rare), Hobby (scarce), Little Ringed Plover, Ringed Plover, Sanderling (scarce), Dunlin, Ruff, Jack Snipe (March/April: scarce), Curlew, Spotted Redshank (scarce), Redshank, Greenshank, Green Sandpiper and Common Sandpiper. From mid-April, Common and Arctic Terns, Turtle Dove, Cuckoo, Yellow Wagtail, Whinchat, Wheatear, Grasshopper Warbler, Sedge Warbler, Reed Warbler, Lesser Whitethroat, Whitethroat, Garden Warbler, Blackcap and Spotted Flycatcher. Maybe scarce migrant. Possible rarity.

*July–September:* Garganey (scarce), Hobby (scarce), Spotted Crake (rare), Little Ringed Plover, Ringed Plover, Grey Plover (rare), Little Stint, Curlew Sandpiper (rare), Dunlin, Ruff, Whimbrel (scarce), Redshank, Greenshank, Green Sandpiper, Wood Sandpiper (scarce), Common Sandpiper, Common and Arctic Terns, roosting Swallows (September) and departing summer visitors. Maybe a rarity.

*October–February:* Wigeon, Teal, Shoveler, Pochard, Goldeneye, Goosander (scarce), Hen Harrier (rare), Jack Snipe (October and November), Woodcock, Long- and Short-eared Owls (scarce), Pied Wagtail (roost in October), Bearded Tit (mostly October and November), flocks of thrushes, finches and buntings including Siskin and Redpoll and roosting Reed and Corn Buntings. Possible rarity.

# RYTON & WAPPENBURY

Map 22
OS map 140

## Habitat

Just 2 miles (3 km) south of Brandon Marsh are two large woods and a disused gravel pit now in the process of reclamation. Within the gravel pit complex, Ryton Pool now forms the central feature of a small picnic area. The pool contains a small island which provides a sanctuary for wildfowl, and extensive tree planting has been undertaken around the perimeter. The remaining gravel workings are being filled with household waste and eventually will be returned to a mixture of agriculture and forestry.

Behind the pool is Ryton Wood. This 168 acre (68 ha) oak-birch wood is now a Warwickshire Nature Conservation Trust (WARNACT) owned reserve. Pedunculate oak standards are dominant, but beneath these are birch, hazel and some old lime coppices. Further south, Wappenbury Wood is a regenerated woodland in the ownership of the Forestry Commission (FC). Here too, oak is the dominant tree. Some parts have a dry, light, sandy soil with a sparse shrub layer and carpet of bracken. Others have a wet, heavy clay, with a dense, impenetrable understorey of hazel, blackthorn and rose.

## Species

There are few outstanding species other than Nightingale, but a good range of woodland birds and a few waterfowl.

At any time of year noisy flocks of Rooks and Jackdaws spar with one another above the canopy, or join the Carrion Crows in mobbing a Sparrowhawk. Down below, the resident Wrens, Dunnocks, Robins, thrushes, tits and finches go about their daily routines. In the more mature woodland, all three woodpeckers and Nuthatch can be found.

Spring is the best time to visit. Then the residents are joined by a good variety of summer visitors. In the scrub around the edges of the woods, Willow Warblers, Whitethroats and Garden Warblers sing, while here and there Tree Pipits nest. From deep within the thick hedges that line the lanes, Lesser Whitethroats can be heard. Inside the woods Chiffchaffs and Turtle Doves sing and Jays call noisily to one another. Suddenly, from deep within a tangle of coppice stools, a Nightingale bursts into rich, liquid song — perhaps to be answered by another on the opposite side of the ride. For this is the Nightingale's stronghold in Warwickshire, with some half-a-dozen singing males most years. The best place to hear them is the damp woodland just where the main footpath enters Wappenbury Wood from the north. They are most vocal from dusk to midnight. Whilst listening, keep an eye overhead for roding Woodcock. As well as Sparrowhawks, Kestrels are regularly present and occasionally a Hobby hunts for dragonflies. In high summer the woods are quiet, but from autumn onwards nomadic parties of tits roam through the sheltered areas and thrushes and finches forsake their daytime pastures for the sanctity of a safe night-time roost. Occasionally in winter a Woodcock is flushed from the bracken along one of the rides.

A few Sand Martins still nest in the sand quarry to the west side of Ryton Wood, although their numbers have diminished of late. Yellow Wagtails are frequently seen around Ryton Pool, where Canada Geese and a pair or two of Mallard or Coot may nest on the island. Little Grebes sometimes nest in the area as well.

In winter a few Tufted Duck are regularly present, while Pochard and Goldeneye appear from time-to-time. One or two Snipe might also be seen, while passage can always bring the unexpected such as a wader or Black Tern.

## Timing

For most song birds early morning in May is the best time to visit, but for Nightingales and Woodcock it is better to try after sunset on a still, warm evening. Song is generally much reduced in cold, wet or windy weather. The winter bird population is often greatest when a covering of snow on the open fields forces birds into the woods to feed.

## Access

Approach from the A423 Coventry to Banbury road. For Ryton Pool turn south-westwards at the roundabout and take the A445 towards Leamington. On the left in ½ mile (0.8 km) is the entrance to the picnic area, where there is a car park and path round the pool.

For Wappenbury Wood continue on the A445 for another ¾ mile (1.3 km), then turn left into Pagets Lane. Continue along this narrow lane until it forks, then take the left fork and park on the verge. Proceed on foot along this private road (public footpath), past Shrubs Lodge and along the footpath into the wood. All of the typical birds can be seen or heard from this public path, so there is no need to wander into the wood.

Ryton Wood is reached by taking an access track south-westwards off the A423 ½ mile (0.8 km) on the Banbury side of the roundabout. There is a small car park at the end of this track. Access is limited to WARNACT members only and is restricted during the shooting season as there are still shooting rights. Visitors are reminded that the WARNACT does not own all of the wood and the boundaries are not yet clearly marked, so make sure you are not trespassing.

## Calendar

*Resident:* Canada Goose, Mallard, Sparrowhawk, Kestrel, Coot, Woodcock, all three woodpeckers, Nuthatch, corvids and other common woodland birds.

*April–June:* Little Grebe, Hobby (scarce), Turtle Dove, Sand Martin, Tree Pipit, Yellow Wagtail, Nightingale and scrub and woodland warblers. Perhaps a passage wader or tern.

*July–September:* Hobby (scarce), Sand Martin and departing summer visitors. Perhaps a passage wader or tern.

*October–March:* Tufted Duck, occasional Pochard and Goldeneye, Snipe, Fieldfare, Redwing, flocks of tits, and thrush and finch roosts.

## Habitat

The 77 acres (31 ha) of Ufton Fields are managed as a nature reserve by the Warwickshire Nature Conservation Trust (WARNACT). Formerly worked for limestone by the cement industry, the site comprises a series of parallel ridges and furrows, some of which hold standing water.

Plants and insects are the primary interests, with several rare or locally scarce calcicoles, especially orchids, but the blend of scrub, carr, grassland and open water is also attractive to birds. As a result of past planting and natural regeneration, much of the centre of the reserve now resembles mature woodland. Regrettably, planting introduced alien pines, poplars and spruce into the area, but the natural regeneration is mostly of oak, ash, alder, willow and hawthorn. There are a number of pools of varying size and depth, some with steeply shelving banks and others showing a gradual gradation from open water, through reed-beds or reedmace to rush and sedge marsh, willow carr and dry scrub.

## Species

Although this site is unlikely to produce very many birds or any great rarities, it is important as prime birdwatching habitats are extremely scarce on the heavy clays of south and east Warwickshire.

The shallow, reedy pools harbour a few pairs of Little Grebe and Tufted Duck, their nests and broods well concealed. Autumn also sees a few Teal and Snipe arrive. They will stay for the winter unless driven out by frost and ice. Pochard are seen less often, but both Grey Heron and Kingfisher make irregular visits to feed. In late summer a Green Sandpiper or other passage wader might pause for a while. Water Rails are present throughout the year and probably breed, though evidence is hard to come by. In summer, squeals, grunts and the seldom-heard song are the only evidence of their presence, but in late autumn and winter, when more birds are present, one might be seen feeding along the muddy margin of reeds. In late summer a small roost of Swallows, House Martins and occasionally wagtails gathers in the reedmace. Sometimes this attracts the attention of a local Sparrowhawk or passing Hobby.

In summer the small bed of common reed holds one or two pairs of Reed Warbler, while in a hard winter it might conceal a stealthy Bittern. Being more catholic in their choice of habitat, Sedge Warblers are more numerous than Reed, nesting in the tangled vegetation and scrub around the pools. Sharing this same habitat are several pairs of Reed Bunting, while in the willow carr the population of Willow Tits has been increased through the provision of special nest-boxes.

Elsewhere on the reserve nest-boxes in the scrub and woodland are regularly used by Tawny Owl, Blue Tit and Great Tit. To a lesser and more variable extent, Tree Sparrows also use nest-boxes, their numbers being greatest in dry summers. Scrub and sylvan warblers like Blackcap, Garden Warbler, Whitethroat, Lesser Whitethroat, Chiffchaff and Willow Warbler are all well represented. Some years ago this was also an excellent site for Grasshopper Warblers, but with the recent widespread decline of this species only one or two are now heard. Sometimes a

Nightingale adds its voice to those of other summer visitors. In autumn and winter there is often a small influx of Woodcock, and thrushes congregate to roost in the blackthorn. Blackbirds and Redwings are most numerous, but Fieldfare and Song Thrush are present too. Very exceptionally a Great Grey Shrike might be seen on top of a thorn bush. Of the specialised feeders, Bullfinches come to feed on bramble and privet, while the alders along the western side hold mixed flocks of tits, including Long-tailed, in late summer and Goldfinch, Siskin and Redpoll from autumn onwards.

The conifers are of less value to birds, but Goldcrest and Coal Tit both feed in them and Great Tits exploit them to some extent. Small numbers of Redpoll may also breed as well, and Woodpigeon, Collared Dove and Turtle Dove all nest here. Where the canopy is sufficiently open to permit reasonable undergrowth, even Chiffchaff and Sedge Warbler will use the trees as song posts. In winter the pines and spruce are used for roosting by finches and buntings. Just occasionally a roosting Long-eared Owl is also discovered. Other birds of prey include Kestrel, which is regularly noted throughout the year, and Sparrowhawk, which normally appears outside the breeding season.

In the grasslands, Skylarks nest and Song Thrushes come from the surrounding hedges and woods to feed on the abundance of snails. There are plenty of anthills too, particularly on the dry ridges, and these attract Green Woodpeckers. Around the meadows, Cuckoos use any convenient perch from which to seek out their hosts' nests, while in one of the old trees a Little Owl might sit dozing in the late-summer sunshine. Occasionally, where young saplings are scattered across the dry ridges, a Tree Pipit might be singing. In spring and autumn one or two Stonechats, Whinchats or Wheatears might pass through, but their occurrences are very variable.

To the south side of the reserve is a rough pasture with rushy, boggy areas that occasionally floods. In winter this sometimes attracts a few wildfowl like Canada Geese or a few Lapwing, Golden Plover or Snipe.

## Timing

An early morning or evening visit in spring and early summer are the best times as birds are then at their most active. High summer is generally poor for birds, but it is the best time for flowers and insects. In autumn and winter, weather is the most important consideration. Avoid wet or windy weather and be prepared for few birds in very frosty or icy conditions.

## Access

Approach from the A425 Leamington to Southam (Daventry) road. At the roundabout in Ufton Village turn southwards into a small lane. Follow the lane round a sharp left-hand bend and onto the following right-hand bend. The entrance to the reserve is on the left-hand side of this bend. Park on the verge, but not on the bend itself. Access to the reserve is restricted to WARNACT members only.

## Calendar

*Resident:* Little Grebe, Tufted Duck, Sparrowhawk (scarce in summer), Kestrel, Water Rail, Lapwing, Little and Tawny Owls, Collared and Turtle Doves, Skylark, Green Woodpecker, tits including Long-tailed and

Willow, Tree Sparrow, finches including Redpoll and Bullfinch, Reed Bunting and other common passerines.

*April–June:* Cuckoo, Tree Pipit, Nightingale (scarce), passage Stonechat, Whinchat or Wheatear, and warblers including Grasshopper, Sedge and Reed.

*July–September:* Grey Heron, Hobby (rare), perhaps a passage wader like Green Sandpiper, Kingfisher, September roost of Swallows, House Martins and wagtails. Passage Stonechat, Whinchat and Wheatear.

*October–March:* Grey Heron, Teal, Pochard, Golden Plover (scarce), Snipe, Woodcock, Kingfisher, roosting thrushes including Redwing and a few Fieldfare, Siskin. Perhaps Long-eared Owl or Great Grey Shrike.

## Habitat

Upton Warren is a Worcestershire Nature Conservation Trust (WNCT) reserve of some 60 acres (24 ha) set in pleasant farming countryside mid-way between Bromsgrove and Droitwich. The reserve consists of two parts, centred respectively on the Moors Pool to the north of the River Salwarpe and the flash pools to the south.

The Moors Pool is the largest and deepest. It has extensive patches of amphibious bistort to provide both food and shelter for birds, and a good margin of sedge, rush and reedmace for nesting. Generally it holds most of the wildfowl. At the northern end is a smaller pool covered with stands of emergent vegetation, mostly reedmace. Separating this from the Moors Pool is a farm track bordered on the south side by a hedge which is always worth checking for unusual migrants. Willows are plentiful and a good stand of mature bankside alders line the Salwarpe.

South of the river, outside the reserve, is an old gravel pit which Hereford–Worcester County Council uses for sail training. It has steep banks and little marginal vegetation and, apart from the resident Canada Geese, attracts only a few gulls and passage terns. Dense, impenetrable scrub flanks the stream leading up the valley towards the flash pools and this, too, is always worth a careful look at migration times.

Highlight of the reserve, though, are the three shallow flash pools themselves. These are the result of subsidence following underground salt extraction and the ground is still sinking. Their natural salinity impedes freezing, except in very cold weather. It also stops vegetation growing over a large area between the pools and this results in an expanse of mud that is a great attraction to waders. A feeding station is maintained at the flashes, which permits close observation of passerines.

## Species

Wader passage is the main interest, but there is a good variety of breeding and wintering wildfowl as well.

Wader numbers are never spectacular, but variety is good and views from the hides are excellent. Spring passage often begins in March with one or two Oystercatcher, Ringed Plover, Dunlin or Redshank, and an increase in the small winter Curlew roost to some 100 birds. Exceptionally they may be joined by a few early Whimbrel, although this species is more likely to be seen in late April or May. Towards the end of March or early in April the first Little Ringed Plovers arrive and soon up to half-a-dozen are running fitfully around the flashes. If conditions are right one or two pairs of Little Ringed Plover and Redshank attempt to breed, but often fail, whereas Lapwing enjoy better breeding success.

April and May see the peak passage, with up to a dozen Ringed Plover and half as many Dunlin and Common Sandpipers. Green Sandpipers, one or two of which may have over-wintered, are also regular in small numbers and a few Ruff usually pass through. Indeed six were recently observed lekking in early May. Most springs bring one or two Greenshank and a resplendent chestnut Black-tailed Godwit, with Sanderling, Spotted Redshank, Wood Sandpiper and Turnstone less often seen. Temminck's Stint, once a regular, has regrettably not appeared in recent springs. With the waders may come a few terns. Mostly these are Common Terns, or Arctic Terns on cool, cloudy days in May, but a party of Black Terns might be seen.

Normally it is July before the first waders begin to move south, but after a poor summer a few may return in June. Green and Common Sandpipers are usually first to show in any number. By late July or early August anything up to 20 Green Sandpipers may have gathered. Counting can be difficult, as they prefer to feed along the flowing stream where they are hidden from view by vegetation.

*Upton Warren is excellent for passage waders such as Green Sandpiper (left) and Common Sandpiper*

Common Sandpipers are less numerous, seldom exceeding a dozen. Similar numbers of Little Ringed Plover, mostly the resident birds, remain until the beginning of September, but the passage of both Ringed Plover and Dunlin is generally weak, with normally no more than three or four birds at any one time. Between late July and early September some half-a-dozen Ruff, up to eight Greenshanks, and one or two Black-tailed Godwits, Whimbrel and Spotted Redshanks are regularly seen. There is also a chance of an Oystercatcher, Wood Sandpiper or Turnstone, while slightly later in the autumn a Little Stint almost invariably appears. In all, over 30 species of wader have been recorded, including Avocet, Pectoral Sandpiper and both Wilson's and Red-necked Phalaropes.

August often brings a Little Gull and perhaps a few terns. Though these are never numerous, they may include a Sandwich or Little Tern. About the same time Snipe begin to congregate around the muddy, rushy margins. By November, 100 or more may be present, accompanied by five or six Jack Snipe. Provided there is no prolonged frost, several Snipe and one or two Jack Snipe usually winter, being joined in March by a few more passage birds. The Jack Snipe are very secretive, however, and are seldom seen. Water Rail are more confiding, with one of several wintering birds quite often feeding in front of a hide.

Sometimes they breed on the reserve. Kingfishers are also resident and can often be seen along the river or around the pools, while Woodcock make occasional visits in bad weather.

Breeding wildfowl include several pairs of Canada Geese, Mallard, Tufted Duck and Coot, and one or two pairs of Little Grebe, Great Crested Grebe and Mute Swan. Both Shelduck and Shoveler have also bred and this is a regular breeding site for Ruddy Duck. Garganey, too, appear quite often, usually on spring and autumn passage, though breeding has occurred in the past.

Post-breeding flocks of wildfowl start to assemble in late summer, when Mallard often reach 300. Their numbers then stay reasonably high until the turn of the year, but decline through the winter. Other species peak somewhat later, with up to 80 Teal and Shoveler in October or November. A handful of Gadwall and perhaps a Pintail can be expected in autumn, followed in winter by a few Wigeon. Shelduck are regular autumn and spring visitors and a herd of Bewick's Swans usually passes through at some time between October and February.

Pochard is the most numerous diving duck, with 100 or more often present and exceptionally twice that number. They may peak as early as November, but January or February are more likely. For much of the winter Tufted Duck do not exceed 50, but April often brings an increase to 70 or so. Both Great Crested Grebe and Ruddy Duck show peaks of around a dozen in early autumn or spring. Other diving duck are scarce and irregular, but one or two Goldeneye or a Goosander appear most years.

Late autumn sometimes brings a scarcer species. Recently Slavonian Grebe, Whooper Swan, two Blue-winged Teal, Ferruginous Duck and Smew have arrived around this time, several of them after storms. Then between February and April there is a chance of a sea-duck like Scaup, Common Scoter or Long-tailed Duck, followed in May perhaps by a Black-necked Grebe. Both Little Owl and Green Woodpecker are resident, often frequenting the orchard behind the flash pools, while Sparrowhawk and Kestrel regularly soar or hover overhead.

Spring migration occurs on a broad front across the valleys of the Severn and Avon and cool, wet weather often produces a 'fall' of Swifts, hirundines and Pied and Yellow Wagtails at Upton Warren. With the wagtails are always a few Whites and sometimes a Blue-headed. Small numbers of Wheatears, Whinchats and Redstarts are also regularly noted, while Water Pipit and Firecrest have been among the more unusual visitors of late. This is not a great place for raptors, but a Marsh Harrier, Osprey or Short-eared Owl might pass through. If cool, overcast conditions coincide with southerly or south-easterly winds, there is a chance of an overshooting migrant. In recent years such conditions have brought Purple Heron, Bluethroat, Savi's Warbler and Golden Oriole.

Yellow Wagtails and Grasshopper Warblers nest among the rushy margins, though the latter is irregular nowadays. A few pairs of Sedge and Reed Warblers also breed and Lesser Whitethroats rattle within the tall hedges and dense thickets of hawthorn and blackthorn. Cuckoos are vocal too, but Spotted Flycatchers are more likely to be seen than heard. From April to September, Hobbies make regular visits, often coming in at dusk in late summer to prey on the Swallows and House Martins.

Departing summer visitors may include a family party of Whinchats or a Redstart. By October the first flocks of Redwing and Fieldfare appear along the hedgerows and parties of Siskin and Redpoll settle into the alders for the winter. A few Tree Sparrows and Linnets are usually among the flocks of finches and buntings and sometimes a Stonechat is around in late autumn. Less often, a Merlin or Short-eared Owl hunts across the reserve, a Rock or Water Pipit feeds around the shoreline, or a Firecrest works its way along a hedgerow. Recently the region's first Yellow-browed Warbler was seen.

## Timing

Disturbance is not a problem, but weather is important. The pools are shallow, so despite their salinity they freeze during very cold weather, forcing wildfowl to move elsewhere. Spring migrants tend to be more numerous in cool, wet weather, and rarities often occur with overcast skies and southerly or south-easterly winds. Most waters need a dry summer before water levels are low enough to expose mud for waders, but at Upton Warren the flash pools are attractive even after a wet season. Autumn vagrants are most likely after a spell of westerly gales. On windy days small birds tend to hide in the thick hedges.

## Access

The reserve lies east of the A38 Birmingham to Worcester road, mid-way between Bromsgrove and Droitwich. It is managed by the WNCT and access is restricted to members only. For the Moors Pool turn off the A38 by the AA telephone box onto a track which leads to a small car park on the left-hand side. Park here and follow the paths to the hides. For the flash pools, park further south in the Hereford-Worcester County Council's sail-training centre (opposite the nursery garden). From here follow the path between the two buildings and alongside the gravel-pit until you see the WNCT sign. Then turn right down the bank, cross the stream and continue to the hides.

## Calendar

*Resident:* Little Grebe, Great Crested Grebe, Mute Swan, Canada Goose, Mallard, Tufted Duck, Ruddy Duck, Sparrowhawk, Kestrel, Water Rail (more in winter), Coot, Lapwing, Little Owl, Kingfisher, Green Woodpecker and common passerines.

*March–June:* Black-necked Grebe (rare), Shelduck, Garganey (scarce), Oystercatcher, Little Ringed Plover, Ringed Plover, Sanderling (scarce), Dunlin, Ruff, Jack Snipe (March/April), Snipe, Black-tailed Godwit, Whimbrel, Curlew, Spotted Redshank (scarce), Redshank, Greenshank, Green Sandpiper, Wood Sandpiper (scarce), Common Sandpiper and Turnstone. From mid-April, Hobby, Common, Arctic and maybe Black Terns, Cuckoo, Swift, hirundines, wagtails including White- and possibly Blue-headed, Redstart, Whinchat, Wheatear, warblers including Sedge, Reed and Grasshopper, and Spotted Flycatcher. Possible migrant raptors or rarity.

*July–September:* Shelduck, Garganey, Hobby, Oystercatcher, Little Ringed Plover, Ringed Plover, Little Stint, Dunlin, Ruff, Snipe, Black-tailed Godwit, Whimbrel, Spotted Redshank, Redshank, Greenshank, Green Sandpiper, Wood Sandpiper (scarce), Common Sandpiper, Turnstone (scarce), Little Gull, terns including perhaps Sandwich and

Little, and departing Swift (scarce September), hirundines, Yellow Wagtail, Redstart, Whinchat, Wheatear and warblers.

*October–February:* Bewick's Swan (scarce), Shelduck (early/late), Wigeon, Gadwall, Teal, Pintail (scarce), Shoveler, Pochard, Goldeneye, Goosander (scarce), Dunlin (scarce), Jack Snipe, Snipe, Woodcock (scarce), Curlew, Rock or Water Pipit (scarce), Stonechat (early or late), Fieldfare, Redwing, Siskin, Redpoll and mixed flocks of passerines including Linnet and Tree Sparrow. Possible sea-duck, raptor or rarity.

# PIPERS HILL &
# TRENCH WOOD

**Map 25**
**OS maps 139 & 150**

## Habitat

Close to Upton Warren are two woods, Pipers Hill and Trench Wood. These woods are typical of many to be found in what was once the Forest of Feckenham.

Pipers Hill, sometimes known as Dodderhill Common, comprises 40 acres (16 ha) of common land standing on a plateau of Keuper marl, which falls away steeply to the west. Within the wood are one or two small pools. The trees are mainly beech, oak and sweet chestnut, many of them very old and therefore attractive to hole-nesting species. Across much of the plateau the heavy shade cast by the lofty beeches precludes all undergrowth save for a patchy cover of bramble, but there is a much better developed shrub layer towards the foot of the west-facing slope.

The 107 acres (43 ha) of Trench Wood are quite different. Some fine, mature oaks still survive from the traditional coppice-with-standards oakwood, but until recently the wood was owned by brush manufacturers who clear felled and replanted large areas with an unusual blend of oak, birch and alder. The Worcestershire Nature Conservation Trust (WNCT) has recently negotiated the acquisition of this wood, which has floristically rich rides as well as a varied birdlife.

## Species

Between them these two woods hold a good range of woodland birds. Pipers Hill is generally the better of the two for arboreal species and Trench Wood for scrub species, particularly Nightingale.

The sparse shrub layer over the plateau of Pipers Hill restricts the breeding community to birds nesting either in the trees or on the ground. In addition to numerous pairs of Blue and Great Tits, the holes and cavities of the larger, older trees hold Little and Tawny Owls and good numbers of Stock Dove, all three woodpeckers and Nuthatch. Treecreepers and Tree Sparrows also breed, although the latter are variable in number. Of the summer visitors, one or two pairs of Redstart nest in holes, while Chiffchaff, Willow Warbler and Wood Warbler nest on the ground. In the more diverse shrub layer of the western slope, common residents such as Wren, Dunnock, Robin and Blackbird are joined in summer by Turtle Dove, Garden Warbler and Blackcap. Spotted Flycatchers occur more sparingly, but a pair or two can usually be found around one of the pools. Winter birdlife is less varied, but the customary parties of tits and finches wander through the wood in search of food. In particular Great Tits, Chaffinches and sometimes Brambling search among the fallen leaves for beechmast.

The common woodland birds of Pipers Hill can also be seen at Trench Wood. In addition, a Sparrowhawk might be glimpsed or a Woodcock flushed from its day-time roost. But summer visitors, especially Nightingales, are of most interest. The district is the Worcestershire stronghold of Nightingales, with a few birds in most woods. In a good year as many as a dozen may be heard singing on a warm spring night in Trench Wood. Two or three pairs of Tree Pipit and

Grasshopper Warbler occur around clearings or young plantations, while Garden Warbler and Blackcap sing within them. Winter brings small feeding flocks of Redpoll and Siskin to the birches and alders, and large numbers of thrushes, including Fieldfare and Redwing, to roost in the dense undergrowth.

Neither wood is noted for unusual species, but a Hobby occasionally passes overhead in summer, and two Firecrests were once discovered at Trench during the autumn.

## Timing

Pipers Hill is a popular common that at times is quite busy. Trench Wood is relatively quiet and undisturbed. Nevertheless, for most song birds in spring and summer it is best to visit as early in the day as possible. The exception of course is Nightingale, which sings best between sunset and midnight on warm, still evenings. Late evening is also a good time to listen for Grasshopper Warblers or to watch Woodcocks roding overhead. As with any woodland, wet and windy weather are best avoided.

## Access

Pipers Hill is crossed by the B4091 Bromsgrove to Alcester road some 2 miles (3 km) south of Stoke Prior. Park on the edge of the wood and explore on foot. Please do not contravene the bye-laws by driving cars too far off the road into the wood.

Trench Wood is best approached from the B4090 Droitwich to Alcester road. At Gallows Green, 2 miles (3 km) east of Droitwich, turn southwards towards Himbleton. Continue along this twisting lane for 3½ miles (5.5 km), past Goosehill Wood and the edge of Himbleton, until you come to the crossroads at Shaftlands Cross. Turn right here and Trench Wood is on the left-hand side in ½ mile (0.8 km). Access to the wood is restricted to WNCT members only.

## Calendar

*Resident:* Sparrowhawk, Woodcock, Stock Dove, Little Owl, Tawny Owl, all three woodpeckers and common woodland passerines including Nuthatch, Treecreeper and Tree Sparrow.

*April–June:* Turtle Dove, Tree Pipit, Nightingale, Redstart (scarce), Grasshopper Warbler, most scrub and woodland warblers including Wood Warbler, and Spotted Flycatcher. Perhaps Hobby.

*July–September:* Quiet in July. Summer visitors leave in August and September. Perhaps Hobby.

*October–March:* Parties of tits and finches, including Siskin, Redpoll and maybe Brambling.

# BREDON HILL &
# THE AVON VALLEY

**Map 26**
**OS map 150**

## Habitat

Bredon Hill is an outlier of the Cotswolds. Its great dome squats in the landscape like some enormous, recumbent animal some 3 miles (5 km) long and 961 ft (293 m) high. Across the flat, fertile Avon Valley the hill dominates the view for miles around.

Neat, picturesque villages nestle into the foot slopes, with the black-and-white, half-timbered cottages of the north contrasting sharply with the mellow Cotswold stone houses of the south. This contrast is also reflected in the landscape. The gentler, southern slopes are often under arable cultivation or grass leys that are grazed by sheep. The northern slopes are too steep for cultivation and are used as rough grazing, with the steeper parts clothed in scrub or mixed woodland. Parklands with their specimen ash, chestnuts and oaks, and one or two stone quarries complete a very diverse habitat.

Immediately north and west of Bredon Hill, the River Avon meanders in broad sweeps through an expansive flood plain. The river is sluggish and fringed with emergent vegetation such as bulrush, bur-reed, and common reed. The humid climate of the valley encourages luxuriant plant growth and an abundance of insects. The banks are lined with a profusion of nettles, meadowsweet, willowherb and pollarded willows. In places there are still remnants of the former osier beds that once were cut for the now defunct basket-making industry. Throughout this stretch the Avon is navigable, with by-pass locks round its weirs that leave small islands of dense vegetation between them and the main river. Many riverside meadows were once Lammas lands, or common hay meadows, enclosed by hawthorn and blackthorn. Few have escaped improvement and even these are no longer managed in the traditional manner. Nevertheless they remain attractive to breeding waders although flash summer floods may wash out many nests. Winter flooding, though, provides ideal feeding conditions for large numbers of wildfowl and waders.

## Species

Bredon Hill is not a place for large numbers of birds. Rather it is a peaceful stretch of countryside in which to seek and enjoy many of the commoner species. There is, none the less, always a chance of something unusual. The Avon Valley is well known for its summer breeding birds and is developing a reputation for its winter wildfowl and waders.

Above the open fields and rough grazing of the summit, singing Skylarks climb ever higher, Lapwings tumble in noisy display and Meadow Pipits go quietly about nesting. In the background there is always the incessant clamour of Rooks, or the sharp chatter of Jackdaws flying to or from their quarry nests. Suddenly, a covey of Grey or Red-legged Partridge breaks cover and heads off on whirring wings. In the parklands the larger holes in gnarled old trees are used by nesting Stock Doves and Jackdaws, and by both Tawny and Little Owls. Quite often a Little Owl sits sunning itself on a bough. Green and Great Spotted Woodpeckers excavate their own holes as well as providing homes for tits and Nuthatch. Around the woodland edge, the periodic crowing or sudden explosive flights of Pheasants are familiar sounds and sights.

The hillside scrub and woodlands hold good numbers of breeding warblers, including Chiffchaff, Willow Warbler, Blackcap, Garden Warbler, Whitethroat and Lesser Whitethroat. Spotted Flycatchers make sorties after insects, as do up to ten pairs of Redstarts that nest on the hill. Here and there a Grasshopper Warbler might be heard singing from a bramble thicket, while among the residents Marsh Tits and Yellowhammers are numerous. Tree Pipits also breed on the hill and a Buzzard or Hobby is often seen overhead. A few Siskin may visit the conifer plantations in spring.

In autumn Bredon Hill is a good place to watch diurnal migration, as small flocks following the Cotswold Scarp pass overhead on their journeys south. Passage begins in September, when Skylarks, Meadow Pipits, Yellow Wagtails, Yellowhammers and finches are on the move. It then reaches its climax in October, when sizeable flocks of Starlings, Fieldfares, Redwings, Chaffinches, Greenfinches and Yellowhammers come from the north and east for the winter. Flocks are largest and most frequent during the first two hours of daylight, and are most evident on still, misty mornings. Some drop down to rest and feed, particularly thrushes which are drawn to the abundance of berries. With them may come a scarce species such as a Black Redstart or passing Hen Harrier.

Winter flocks of Chaffinches, Greenfinches and Yellowhammers gather to feed among stubble or in sheltered furrows. During frosty weather they also congregate around sheep feeding stations, where they are liable to attract the attention of a resident Sparrowhawk. In some years they are joined by good numbers of Brambling. Large flocks of Rooks, Jackdaws and thrushes are also a familiar part of the winter scene, feeding in the fields by day and roosting in the woods at night.

As the days lengthen, so Skylarks, Lapwings and Meadow Pipits return to their territories. Spring sees an altogether swifter and less evident passage than autumn and before long the hill is resounding once again to the songs of innumerable summer visitors. A handful of passage Wheatears may pause to refuel in the summit fields and there is always a slim chance of a Nightingale or something really exciting like a Firecrest or Dotterel.

Down in the Avon Valley is a very different world. Here the stands of emergent vegetation along the river form breeding sites for many pairs of Moorhen and a few Coot. They also shelter many Mallard ducklings that have hatched in the crowns of pollarded willows. One or two Mute Swans glide majestically along, up-ending in search of food, but this and many other species have declined dramatically as boat traffic has increased. Grey Wagtails frequent the locks and weirs and the blue flash of a passing Kingfisher is occasionally seen, although most now prefer to breed along the quieter tributaries than the main river.

The luxuriant vegetation along the riverbanks holds many warblers. Sedge Warblers are well distributed and their scratchy song might be heard from almost any patch of willow or thorn scrub. Reed Warblers are confined more to the patches of common reed, but being colonial several may be heard singing together.

Most birdwatchers hope to see and hear the Marsh Warbler, which is the real speciality of the valley. Worcestershire, with up to 50 pairs, used

*The Avon Valley was formerly a stronghold of the rare Marsh Warbler, but numbers have recently collapsed*

to hold three-quarters of the British population of this very rare warbler, but this has recently collapsed to the point where perhaps fewer than half-a-dozen pairs survive. The cause may be changes at the wintering rather than the breeding grounds, but it does mean we are unable to disclose the precise whereabouts of any remaining pairs. If birds are in the area they should be visible and audible from the footpaths described, but you will need to be patient. Do not expect just to turn up and be lucky. Above all do not trespass into sensitive habitats where you can do considerable damage searching for a bird that may not even be there. Given a chance by birdwatchers and a lot of help by nature, there is just a hope that the population of this delightful songster, with its amazing gift for mimicry, might yet recover.

Snipe, Redshank, Yellow Wagtail and Grasshopper Warbler all breed sparingly in the damper unimproved meadows, with perhaps an occasional pair of Curlew too. Swallows and House Martins come regularly to the river to feed throughout the summer and Cuckoos are widespread. In spring there is always a chance that a rarity might follow the river, though few will be as exciting as a recent Black Kite.

More intensive winter watching in recent years has revealed some large concentrations of wildfowl and waders. Of most interest is the herd of Bewick's Swans along the Worcestershire/Gloucestershire border just north of Tewkesbury. Swans were first reported here in the 1980/1 winter and the herd has subsequently grown to well over 200. Occasionally it is joined by a Whooper Swan or a small party of White-fronted Geese. Slimbridge is only 22 miles (35 km) away and the swans resort there in the hardest weather, returning again once a thaw sets in. With 20 or so Mute Swans on the river as well, there is a chance of seeing all three species of swan at the same time.

Some half-a-dozen Cormorants are regularly on the river, one or two Little Grebes can usually be seen and Kingfishers pass up and down. Ducks are often present too. Most likely is a small flock of Wigeon, which for part of the winter at least grazes the lush meadows and roosts on the river. During very cold weather numbers may swell to as many as 300, or exceptionally well over 1,000 in March. With them may be a few Gadwall. Up to 50 Mallard and 250 Teal are also likely, and one or two Shoveler or even an elegant Pintail are possible. Diving duck congregate on the river as well, especially when still waters are frozen solid. Pochard and Tufted Duck are most probable, but a few Goldeneye and Goosander are quite possible, especially in February and March. There may even be a Smew, though the 14 that gathered there early in 1985 are unlikely to be matched for a long time.

Large flocks of Lapwings, 1,000 or more strong, assemble on the adjoining fields, often accompanied by a few Golden Plover. As floodwaters subside, the wet meadows attract plenty of Snipe and Dunlin, a few Redshank and maybe a wintering Ruff or Bar-tailed Godwit. The hard weather early in 1985 brought unprecedented numbers, with 6,000 Lapwing, 2,000 Golden Plover and 600 Dunlin. Large flocks of gulls and thrushes also visit the flooded fields, with up to 2,000 Black-headed Gulls and 100 Common Gulls, the latter especially in spring. If the meadows are still wet at this time, then passage waders may include Ringed Plover, Grey Plover, Black-tailed Godwit, Whimbrel, Greenshank and Green and Common Sandpipers as well as those already mentioned. A Little Gull is even possible, while Yellow Wagtails are regular visitors. With them may be a few White Wagtails or a Blue-headed. Such concentrations invariably attract predators. Sparrow-hawks are seen daily and dashing, twisting Merlins are quite regular in late winter. Less often a majestic Peregrine stoops at breath-taking speed into a flock of roosting birds.

## Timing

Both Bredon Hill and the Avon Valley are popular, especially at weekends and summer Bank Holidays, and therefore best visited in the early morning, when they are quietest. Misty autumn mornings immediately after day-break are best for diurnal migration.

The river is fished and carries a lot of pleasure craft, while in places sailing adds to the disturbance. Winter is quieter and weather then becomes the key factor. Numbers are best in hard weather, but if the ground is frozen solid or covered in snow there may be little to see. Equally birds are scarce in mild weather. Flooding occurs after heavy rain or rapid thaws, but receding floodwaters are often most productive.

## Access

Bredon Hill lies between the A435/A438 Evesham to Tewkesbury, B4080 Tewkesbury to Pershore and A44 Pershore to Evesham roads. The summit can be reached by a short, sharp climb of nearly 1 mile (1.5 km) from Elmley Castle, a long, steady climb of nearly 2 miles (3 km) from Kemerton or a steep and tortuous climb of 2½ miles (4 km) from Bredon's Norton. In addition, the Wychavon Way crosses the eastern end of the hill, between Ashton-under-Hill and Elmley Castle, and there are numerous other public footpaths. Parking is difficult in the narrow village streets, so please take care and avoid blocking access ways.

The best stretch of the River Avon is the 14 miles (23 km) between Pershore and Tewkesbury. In summer the stretch upstream of Strensham

Lock is generally of most interest, whereas in winter the stretch downstream is usually the better. Access points are:

*Nafford* Leave the A4104 Pershore to Upton-upon-Severn road southwards on the B4080 towards Tewkesbury. Turn left at the first cross-roads in Eckington village and continue for 1 mile (1.6 km), where a small car park overlooks the river on the left. Park here, walk a little further along the lane and take the footpath on your left down to the lock. Cross onto the island to view the river. Sedge and Reed Warblers can usually be seen on the island, but please do not leave the footpath.

*Eckington* Proceed as above from the A4104, but immediately after crossing the old sandstone river bridge park in the small car park on the left-hand side. Cross over the road and take the footpath westwards along the south bank of the river. This can be followed through to Strensham Lock, a distance of 2 miles (3 km), or on to Bredon, a distance of 4 miles (6.5 km).

*Twyning Green* At the M50/A38 junction take the lane southwards to Twyning Green. Park at the end and follow the footpath southwards along the Gloucestershire bank. This continues for just over 2 miles (3.5 km) until it rejoins the A38 on the edge of Tewkesbury. Winter wildfowl and waders usually occur along this stretch and in the vicinity of Bredon.

## Calendar

*Resident:* Little Grebe, Mute Swan, Mallard, Buzzard, Sparrowhawk, Kestrel, Red-legged and Grey Partridges, Pheasant, Moorhen, Coot, Lapwing, Snipe, Woodcock, Curlew (scarce), Redshank, Stock Dove, Little Owl, Tawny Owl, Kingfisher, Green and Great Spotted Woodpeckers, Skylark, Meadow Pipit, Grey Wagtail, Marsh Tit, Nuthatch, Jackdaw, Rook and other common corvids, tits, finches and buntings.

*March–June:* Lapwing, Skylark and Meadow Pipit return to hill in March. Ringed Plover, Grey Plover (rare), Dunlin, Ruff, Black-tailed Godwit (scarce), Whimbrel (scarce), Greenshank (scarce), Green Sandpiper, Common Sandpiper and gulls including Common and maybe Little. From mid-April, Hobby (scarce), Cuckoo, Swallow, Tree Pipit, Yellow Wagtail, White Wagtail (scarce), Redstart, Wheatear, Grasshopper Warbler, Sedge Warbler, Marsh Warbler (from late May: scarce), Reed Warbler, Lesser Whitethroat, Whitethroat, Garden Warbler, Blackcap, Chiffchaff, Willow Warbler, Spotted Flycatcher and maybe Siskin. Possible rarity.

*July–September:* Many summer visitors leave during August. September; Skylark, Meadow Pipit, Yellow Wagtail, Yellowhammer and finches move south-westwards.

*October–February:* October; Skylark, Meadow Pipit, Fieldfare, Redwing, Starling, Chaffinch, Greenfinch, Yellowhammer and Brambling move south-westwards. Maybe a rarity. Winter; Cormorant, Bewick's Swan, Whooper Swan (scarce), White-fronted Goose (scarce), Wigeon, Gadwall (scarce), Teal, Pintail (rare), Shoveler (scarce), Pochard,

Tufted Duck, Goldeneye and Goosander (scarce), Smew (rare), Merlin (scarce), Peregrine (rare), Lapwing, Golden Plover, Dunlin, Ruff (scarce), Bar-tailed Godwit (rare) and gulls.

# THE MALVERN HILLS

Map 27
OS map 150

## Habitat

The narrow, spectacular switchback of the Malverns rises abruptly to reach a height of 1,394 ft (425 m) on Worcestershire Beacon and dominate the skyline for miles in every direction. To the east is the broad expanse of the Severn plain and to the west the rolling red fields of Herefordshire. The ridge, which is 8 miles (13 km) long, is formed by some of the oldest rocks in England and is managed by the Malvern Hills Conservators. From its open summits stretches a panorama of wooded slopes, commons, parklands, orchards and rich, fertile farmland studded with tiny villages. Nestling right into the foot of the hills is the elegant spa town of Great Malvern.

The summits have short grass swards with patches of heather and bilberry. These grade quickly into hillsides of bracken and bramble, with the lower slopes invaded by gorse, broom, thorn scrub, rowan and light birch woodland. Lower still, in the more sheltered valleys and foothills, the scrub develops into broad-leaved woodlands of oak and ash above birch, hazel and occasionally yew. In autumn, Happy Valley, with its rowans, is one of the best migrant 'traps' in the region. Elsewhere there are some fine old orchards, with gnarled, lichen-clad fruit trees that are favoured by hole-nesters. Several old stone quarries have vertical cliffs and ledges which are being recolonised by gorse, bramble, wild rose and saplings of ash, elder, sycamore and willow. Finally, a small reservoir and one or two tiny pools complete the mosaic of habitats.

Castlemorton is one of many commons around the foothills. Here, despite sheep grazing, dense clumps of gorse and bramble are invading the rough, low-lying permanent pastures. Traversing the middle of the Common is a small stream, which is flanked on either side by marshy areas of sedge and rush. In summer these dry out completely, but in winter they are waterlogged if not flooded.

Inevitably, the public throng to the hills and commons, especially on fine, sunny days in summer, and in places they are causing serious erosion. However, despite the ramblers, picnickers, horse riders and hang-gliders, there are still plenty of places that are sufficiently quiet to hold a good range of interesting birds.

To the north, the back-bone of the Malverns subsides into a tail of hilly wooded country dissected by fast-flowing streams. Amidst this delightful, tranquil countryside is the 60 acres (24 ha) reserve of the Knapp and Papermill. This reserve, owned by the Worcestershire Nature Conservation Trust (WNCT), lies astride a fault, which brings Triassic marls and sandstones up against Silurian limestones resulting in a particularly rich flora. The central feature of the reserve is the narrow, secluded valley of the Leigh Brook with its woodland, old meadows and orchards. The steeper slopes are clothed in broad-leaved woods which, from the presence of small-leaved lime and wild servicetree, are believed to have shaded the valley for many thousands of years. Today their standard oaks rise above coppices of hazel and small-leaved lime. On the gentler slopes, the unimproved meadows are full of wild flowers, which in high summer attract countless insects.

The brook, which is lined by coppiced alders and pollarded willows, has both fast-flowing shallows and deeper, slower reaches with high banks.

## Species

The Malverns are an area for birds of grasslands, commons, woods and streams.

Skylarks and Meadow Pipits are the characteristic species of the open summits. In spring and summer the air is full of their songs. In autumn both species move down to the lower pastures and commons for the winter. Much scarcer, but equally at home on the short summit swards, are Wheatears. As many as six pairs have nested in the past, though one or two are all that can be expected these days. They can often be seen in the vicinity of British Camp, with many more passing through in spring and autumn.

*One or two pairs of Stonechat nest on the Malvern Hills*

Scolding Wrens flit across the bracken-clad slopes and in good years a Grasshopper Warbler reels from a patch of bramble. Clumps of gorse or thorn scrub might hold a pair of Stonechats, if a hard winter has not depleted the population. More often, though, a few pass through each spring and autumn, visiting the commons as well as the hills. Gorse, bramble and thorn scrub are also breeding haunts of good numbers of Whitethroat, Linnet and Yellowhammer, while Tree Pipits occur sparingly where there are scattered trees or light woodland. Anthills are common amongst the short turf of the hills, so Green Woodpeckers are plentiful and often one rises from the ground ahead and makes off in bounding flight. Frequently they choose old orchard trees for their nests – a niche which they share with Lesser Spotted Woodpeckers and a few pairs of Redstart.

The deciduous woods around Malvern hold a good range of typical species. Warblers include Blackcap, Garden Warbler, Chiffchaff, Willow Warbler and a few Wood Warblers. The latter are localised, but their silvery trills can usually be heard in beechwoods on Midsummer Hill or in the woods above Wyche Quarry. Other regular summer visitors are Redstart, Spotted Flycatcher and, very occasionally, Pied Flycatcher. Of the resident species, Jays are plentiful and Great Spotted Woodpecker,

Nuthatch and Treecreeper can all be found. Jackdaws circle high above the quarries and down beneath them small birds like Yellowhammer nest in the herbage and developing scrub.

Autumn usually brings a good variety of passage birds to the hills. Once the summer visitors have left, flocks of thrushes and finches, including Fieldfare, Redwing and Mistle Thrush, congregate in Happy Valley to feed on the abundance of rowan berries and other fruits. With them are normally a few late Ring Ouzels, with perhaps a dozen or more on occasions. Later, a rarity like Snow Bunting might appear with the flocks of larks, pipits and thrushes that are moving south. Indeed, Snow Buntings are seen quite often on the summits in November, sometimes in small parties, and it is possible that a few regularly pass through undetected. Autumn also brings large parties of Siskin and Redpoll to birches and alders, while yew and ivy berries may sustain an over-wintering Blackcap.

Of the raptors, Kestrels are commonest with sometimes a family party hanging into an updraught. Sparrowhawk and Buzzard are also frequent visitors, especially in the more wooded country to the west and north. Even Ravens make sporadic forays from their more traditional haunts further west. Autumn often brings a migrant Hobby, perhaps moving south with the departing Swallows and House Martins. Even a Peregrine could be seen, though the sight of one being mobbed by two Hobbies is unlikely to be repeated.

On the commons, patches of gorse and bramble hold good numbers of Linnets and Yellowhammers, a few pairs of Whitethroat, and maybe a Grasshopper Warbler or two. Both Stonechat and Whinchat occur on passage, when they perch prominently on the top of gorse bushes and bramble thickets. In winter Redwing and Fieldfare strip the berries from the hawthorn and mixed flocks of finches and buntings become the target of Sparrowhawks. Sometimes a Great Grey Shrike makes a rare appearance. Along the stream a few Snipe probe the mud between clumps of rush and sedge, accompanied perhaps by a Jack Snipe or Water Rail. Very exceptionally, other waders may drop in on passage.

The woodlands of the Knapp and Papermill reserve hold similar species to those around Malvern, with the exception of Wood Warbler, but the meadows and stream add more diversity. Both Dipper and Kingfisher nest along this short stretch of brook, and all three wagtails breed in the vicinity. Winter sees flocks of Siskin and Redpoll in the alders and thrushes in the meadows and woods, while at dusk a Little Owl, or rarely a Barn Owl, goes hunting.

## Timing

At weekends and Bank Holidays it is best to start early on the hills and commons, before too many people get there. For the Knapp-Papermill this matters less, though song birds, of course, are more vocal in the mornings and evenings anyway. In very wet or windy weather you can expect to see very little on the hills. Fine sunny days are best for soaring raptors.

## Access

The northern foothills are encircled by the A449 Worcester to Monmouth and B4232 roads. From these there are innumerable access points onto the hills. Both the A449 and the B4218 cut through the hills *en route* from Great Malvern to Ledbury. South of Little Malvern the hills are more remote and less disturbed. For Castlemorton Common leave the A4104 Little Malvern to Upton-on-Severn road southwards at the crossroads in Welland onto the B4208 towards Gloucester. Just under 1 mile (1.5 km) turn right onto a minor road that crosses the common, park and explore on foot. Then continue to the foothills and park where the lane turns sharp left for access to the hills themselves. Midsummer Hill and The Gullet usually repay exploration. The lane then continues through to join the A438 Tewkesbury to Ledbury road from the north at Hollybush.

The Knapp-Papermill reserve is best approached from the A4103 Worcester to Hereford road. Leave this road to the west by the bend in Bransford village. Take the left fork towards Alfrick Pound at the next junction and continue for 3 miles (5 km) along a twisty lane until you cross the bridge over Leigh Brook. Park on the verge by the bridge, walk up the hill and enter the reserve on the left through the garden of 'The Knapp' house. Access is restricted to members of WNCT, from whom further details can be obtained.

# Calendar

*Resident:* Sparrowhawk, Kestrel, Little Owl, Tawny Owl, Kingfisher, all three woodpeckers, Skylark, Meadow Pipit, Grey Wagtail, Dipper, common woodland species including Nuthatch and Treecreeper, Linnet and Yellowhammer.

*April–June:* Buzzard (scarce), Tree Pipit, Yellow Wagtail, Redstart, Whinchat, Stonechat, Wheatear, scrub and woodland warblers including Wood Warbler, Spotted Flycatcher and Raven (scarce).

*July–September:* Quiet in July. Summer visitors depart in August and September. Buzzard (scarce), Ring Ouzel (September) and Raven (scarce). Perhaps passing raptors.

*October–March:* Buzzard (scarce), Water Rail (rare), Jack Snipe (scarce), Snipe, Barn Owl (rare), Stonechat (early and late), Ring Ouzel (October), Fieldfare, Redwing, Raven (scarce), Siskin, Redpoll and Snow Bunting (November, scarce). Perhaps passing raptors.

# THE WYRE FOREST

Map 28
OS map 138

## Habitat

The Wyre Forest stands on a plateau immediately west of the River Severn, just upstream of Bewdley. Underlain by sandstones, marls, conglomerates and limestone bands, this remnant of ancient hunting forest contains an outstanding flora and fauna within its 6,000 acres (2,400 ha). This is not only the largest and most diverse woodland in the West Midlands, but one of the best remaining native woodlands in Britain. About a fifth is a National Nature Reserve (NNR) and the West Midland Bird Club (WMBC) and the Worcestershire Nature Conservation Trust (WNCT) have small reserves too. Just over half, though, belongs to the Forestry Commission (FC) and since 1927 this has been progressively converted into conifer plantations. Now that the older ones have matured, clear felling and replanting has begun.

The woods of Wyre are a unique blend. The oakwoods on the acidic plateau soils recall the coppices of Wales, whereas those on pockets of clay are reminiscent of the mixed woods of East Anglia and the richer valley woods are similar to those of the southern Welsh borders. Indeed, several plants and animals here are on the very edge of their range.

The glory of the forest is its oakwoods, which are developing the closed canopy of high forest now that coppicing has virtually ceased. Sessile is the commoner, but pedunculate also occurs and many trees show intermediate characteristics. On the plateau and steep valley sides, where the soils are acidic, the oak generally occurs with birch above an understorey of heather, bilberry, bracken and great wood-rush. In some parts a few stools are still coppiced to yield timber for rustic furniture and bark for tanning. In others, trees have been thinned to leave standards that overshadow birch scrub, bramble and bracken, together with a few holly and yew.

The pockets of calcareous clay, derived from limestone bands, support a very different woodland. Here ash flourishes along with hazel and dogwood, over a field layer of dog's mercury and primrose. The most diverse, though, are the valley woods, which can be seen at their best along the beautiful Dowles Valley. This is the gem of the forest, with a brook that tumbles and twists over boulders and pebbles. Above its high, overhanging banks, old orchards and tiny meadows nestle into the narrow valley. Towering above these are steep slopes of superb hanging oaks that rise to merge with those on the plateau.

The valley flora is extremely varied, with alder, ash, elm, small-leaved lime and willow as well as oak. Beneath these is a shrub layer rich in autumn berries, with rowan, blackthorn, hawthorn, guelder rose and wild rose. In the damp flushes, both pendulous sedge and meadowsweet thrive. The less acidic soils were cleared of trees to make space for grazing meadows and orchards. Now neglected, the gnarled, twisted trees of apple, damson and plum provide a wealth of nest-sites, while the meadows are alive with harebell, yellow rattle, cowslip, green-winged orchid, meadow saffron and a host of butterflies.

The Wyre is also renowned for the multitude and variety of its insects. These range from an abundance of wood ants to conspicuous dragonflies and butterflies like high brown and silver-washed fritillaries.

Two are national rarities, namely Kentish Glory and Alder Kitten Moth. This wealth of insects helps sustain a good population of birds.

Space precludes reference to every nook and cranny of this fascinating forest, but mention must be made of Seckley Wood. Forming the north-eastern extremity of the forest, this is a mixed wood with oak and birch on the predominantly light, sandy soils and alder on the wetter clays. There are some fine old beeches too, and a few old cherry orchards nearby. The plateau ends abruptly in a spectacular sandstone cliff in which pockets of clay cause occasional landslides. From the top the view is superb, with the Severn gliding beneath, while on the far bank steam trains on the Severn Valley Railway puff slowly past the tiny Trimpley Reservoir.

The FC plantations make a sharp contrast to this idyllic scene. Here lofty spruce, pine, larch and Douglas fir stand in serried ranks. Shafts of sunlight along the forest rides admit a limited ground flora, but inside the plantations are dark and largely lifeless. In the New Parks, though, conifers were used only to replace impoverished oakwoods on the higher ground. Lower down, on the more fertile soil, the better oaks were kept and underplanted with beech. Belts of oak were also left alongside roads and tracks. In fact, the FC policy is to maintain hardwoods over a third of its Wyre woodlands, which obviously benefits the birds. As the woodlands have matured, even the plantations have become more attractive and today they hold most birds typical of coniferous woods.

## Species

Few forests can boast a better avifauna than the Wyre.

Breeding activity begins quite early in the spring, long before the leaf-buds have burst. Then, on bright, crisp mornings, residents like Robin, Dunnock, Wren, Blackbird, Song and Mistle Thrushes, and Chaffinch gradually commence singing in defence of their territories. At first no more than a spasmodic snatch may be heard, but before long they will all be in good voice and reinforced by Blue and Great Tits and Treecreeper. The ringing calls of Nuthatch, a Green Woodpecker's yaffle and the vibrating echoes of drumming Great and Lesser Spotted Woodpeckers all add to the chorus.

Each of these birds has its own particular habitat preference. Nuthatches keep to the older, broad-leaved trees, especially beech. Green Woodpeckers search the wood, clearings and meadows for wood ants, but often nest in a birch or orchard tree. Most widely distributed is the Great Spotted Woodpecker, while the tiny Lesser Spotted is the scarcest and hardest to find, particularly when it is feeding high in the oak canopy. In the old orchards, however, it is much more easily observed as it flutters from branch to branch.

Once the summer visitors join the resident songsters, the forest reverberates to a symphony of bird song. First to arrive are Chiffchaffs, whose metronomic notes are sometimes heard as early as mid-March. Most are found along the luxuriant valleys, but a few sing along forest rides. Next to arrive are Willow Warblers, which quickly become ubiquitous. The trickle of migrants steadily increases to a crescendo in May. Cuckoos call across the valleys and along Dowles Brook both Blackcap and Garden Warbler add their sustained, harmonious melodies. Both prefer a good shrub layer, but whereas Blackcaps are found under a closed canopy, Garden Warblers usually nest beneath an open sky and are consequently the scarcer of the two. Clearings or open glades are also the haunt of Whitethroats and Grasshopper Warblers, which nest sparingly in the undergrowth. A few pairs of Tree Pipit also breed, with the abandoned railway line a good place to look. Other migrants are Turtle Dove and Spotted Flycatcher, with several pairs of the latter along the streams and in the orchards.

The specialities of the Wyre are that characteristic trio of sessile oakwoods, Wood Warbler, Redstart and Pied Flycatcher. Wood Warblers occur wherever the shrub layer is sparse, but the field layer good, with some 20 pairs along Dowles Brook alone. They are relatively easy to watch as they deliver their shivering song from a low perch. Redstarts are fewer and mostly confined to valleys like that of the Dowles, where they occupy nest-boxes or holes in oaks and contorted orchard trees. Frequently high in the oak canopy, they can be surprisingly inconspicuous, but during courtship chases early in the season the constant flicking of their bright fiery tails can hardly fail to be noticed.

*The Pied Flycatcher is the most captivating of the Wyre Forest birds*

The most captivating of all the Wyre birds is the Pied Flycatcher. It seldom arrives until the end of April, by which time most suitable nest-holes are already occupied. In the past this may have restricted breeding, but now in the Wyre, as elsewhere in the country, the population has expanded with the provision of nest-boxes and there are currently at least a dozen pairs. The stream, overhanging oaks, old orchard trees and open meadows are ideally suited to its needs and pairs can now be seen all along the Dowles Brook, but especially in the Fred Dale Reserve, in Knowles Coppice and around Knowles Mill. When they first arrive, the smart black-and-white males are especially active. Once the nestlings have fledged though, they disappear along with many other species into the now leafy oaks, where they are extremely hard to see. Indeed, birdwatching in the forest in high summer can be singularly unrewarding.

Sit quietly by the brook, though, and riparian species can be seen at any time. On the deeper, sluggish stretches of the Dowles, Kingfishers nest in the high, sandy banks. Many views are a quick blurr of brilliant blue skimming just above the water, but sometimes one is seen on an overhanging perch when its brilliant turquoise and chestnut plumage can be really appreciated. On the faster-flowing reaches, Kingfishers are replaced by Dippers, which nest under the overhanging banks and bridges. Often one is seen bobbing nervously on a favourite boulder or flying fast and direct just above the stream. Sharing their haunts are Grey Wagtails, which dart after insects with tails pumping.

The most elusive bird is the Hawfinch. Precisely how many inhabit the Forest is unknown, but parties up to 20 are seen from time-to-time, usually feeding beneath beeches in early spring. Once the leaves open detection is difficult, unless you know their sharp, metallic 'tzik' call, in which case you might find them, in mid-summer, feeding in the canopy on the larvae of the oak tortrix moth. They nest in loose colonies, sometimes in Douglas firs.

The conifer plantations of the Wyre hold much more beside the usual Goldcrests and Coal Tits. Siskin flocks of 200 or more gather in the spruce plantations every March and April, and breeding has been suspected more than once. Crossbills too are now regularly present and parties up to 50 are quite common between January and March, with small numbers breeding in some years. Surprisingly few occur in late summer and early autumn, though, which is the customary time for irruptions. Firecrests have also been seen several times, mostly in spring, and breeding remains a possibility. Douglas firs have again been favoured, but finding them in such a vast area is a big problem. The same is true of Long-eared Owls, a pair or two of which inhabit the vast plantations.

Calm, sunny mornings are best for raptors, when they rise into developing thermals. Sparrowhawks and Kestrels are common, but Buzzards and Goshawks are less regular although they do appear from time-to-time, as indeed do Ravens. On rare occasions an Osprey might even be seen drifting high above the Severn. At twilight Little Owls emerge from their nests in orchards or around the forest edge, while hooting Tawny Owls are plentiful in many parts of the forest.

In winter, parties of small birds wander along sheltered valleys like the Dowles in search of food. Tits are the most evident, but Goldcrests, Treecreepers and even a Lesser Spotted Woodpecker may be loosely associated with their nomadic flocks. Great Tits and several hundred Chaffinches come to feast on beechmast, where they are sometimes joined by a few Brambling. The latter may be more numerous in early spring, though, when as many as 100 gather in conifers, especially larch. Flocks of Redpoll and Siskin occur from autumn through to spring. At first they take seeds from birch catkins, but later they move into streamside alders. Fieldfare, Redwing and Blackbird are also numerous, feeding away from the Forest, but coming in to roost at night. The resident population of Woodcock is also swollen by winter immigrants from the Continent.

Of the rarer species, Hoopoe and Golden Oriole have occurred in spring, and Great Grey Shrike in winter. Mandarin and Wood Duck sometimes nest along the Dowles, where a Black Stork once appeared. Whilst in the Forest do not forget the Severn and the nearby Trimpley Reservoir. Cormorants often move up and down the river outside the breeding season, pausing *en route* to feed on the reservoir. In winter, Goldeneye and Goosander do likewise, while spring and autumn might bring Shelduck or a party of Red-breasted Mergansers. Even a vagrant diver, Shag or wild swan is a possibility.

## Timing

Although a popular spot, the Forest is seldom too crowded and few people walk very far from their cars. In spring and summer birds are most active early in the morning and late in the day. For resident species try March or early April, before the trees are in leaf. This is certainly the best time for Hawfinches as they are then more active and noisy. For summer visitors, it is better to visit in May, when numbers are highest. Raptors become more active by late morning as the ground warms up and thermals begin to rise. March and April are usually the best months for birds of prey. By high summer most birds are hard to find as they retreat to the leafy canopy with their newly-fledged young. Timing matters less in winter as birds feed throughout the short days, especially during very cold spells. Occasionally the FC woodlands are used for car rallies or orienteering.

## Access

The Forest is best approached from the A456 Bewdley to Ludlow or B4194 Bewdley to Kinlet roads. The best access points are as follows:

*Dowles Brook* Leave Bewdley on the B4194 towards Kinlet. In a little under 1 mile (1.5 km) the road passes over Dowles Brook and between and old railway embankment, where there is limited parking on the left. Follow the footpath westwards alongside the old railway line. This takes you through the Fred Dale Reserve (WMBC and WNCT reserve), crosses over the Brook and eventually comes to Dry Mill Lane. Continue to follow this lane alongside the Brook into the heart of the Forest. After 1 mile (1.6 km) the lane passes Knowles Mill, an old stone house on the opposite bank. There is a footbridge across to the house which also leads onto a footpath up through Knowles Coppice (WNCT reserve) to the old railway line. A pleasant round walk can be made by taking this path and then returning along the old railway.

*Seckley Wood* Continue past Dowles Brook towards Kinlet for a further 1½ miles (2.5 km) and park in the FC picnic site. There are three waymarked forest trails from here. Take the longer (red) trail, but divert from it, past the Seckley Beech and round the forest track that overlooks the River Severn and Trimpley Reservoir. A telescope is needed to view the latter.

*New Parks* This is the main area of conifer plantations. Leave Bewdley on the A456 towards Ludlow and continue for 2½ miles (4 km) until you come to the FC Callow Hill car park and Visitor Centre on the right. Park here and explore one of the three waymarked trails. Again, the longer (red) route is usually the best; it is 3½ miles (6 km) long. From this trail one of two footpaths lead straight on into the Dowles Brook.

## Calendar

*Resident:* Sparrowhawk, Kestrel, Woodcock, Stock Dove, Little Owl, Tawny Owl, Long-eared Owl, Kingfisher, all three woodpeckers, Grey

Wagtail (scarce in winter), Dipper, Nuthatch, Treecreeper, Siskin, Redpoll, Hawfinch and other common woodland passerines.

*April–June:* Winter visitors depart in April. Mandarin (scarce), Turtle Dove, Cuckoo, Tree Pipit, Redstart, Grasshopper Warbler, Whitethroat, common woodland warblers including Wood Warbler, Firecrest (rare), Spotted Flycatcher and Pied Flycatcher. Maybe Buzzard, Goshawk, Raven or Crossbill. A few wildfowl along the Severn. Possible rarity.

*July–September:* Quiet in July. Summer visitors leave in August and September, when maybe passage raptor. Perhaps irruption of Crossbills.

*October–March:* Roosting thrushes especially Redwing, flocks of tits and finches including Brambling (scarce) and Crossbill (especially January-March). Siskin and Brambling in conifers during March. Possible raptor or Great Grey Shrike. Maybe Cormorant, Goldeneye, Goosander or vagrant along river.

## Habitat

Woolhope is a small village 7 miles (11 km) east-south-east of Hereford. It lies at the centre of some beautiful, rolling countryside where rich, fertile vales are enclosed by limestone hills, each surmounted by a continuous stretch of superb woodland. There are one or two special spots for wildlife, but generally this is an area to be explored at leisure from the many roads and footpaths that traverse it.

The largest woodland is the Forestry Commission (FC) Haugh Wood, where the plateau and steep hillsides are covered with plantations of pine and larch that are fringed with beech. Parts have more extensive stands of beech, while wide sweeps of bracken cover the recently felled areas. Through the centre is the National Trust (NT) Poor Men's Piece, a narrow strip of rough oakwood with some wild service trees. Immediately east of Haugh Wood is Broadmoor Common, a tract of heath with scattered trees, thickets and extensive patches of gorse.

Further south are two small, but interesting woodland reserves. At Nupend the Herefordshire and Radnorshire Nature Trust (HRNT) owns 12 acres (5 ha) on a steep ridge of Silurian limestone clothed to the west by dense yew and to the east by coppiced oak. There is also some ash and beech, especially to the south. The yew excludes all undergrowth and little but moss grows beneath the oaks on the dry ridge, but on the lower slopes field maple, hazel and some large wild service trees all thrive. The yews especially are visited by many birds in autumn.

The HRNT owns a further 22 acres (9 ha) in Lea and Paget's Woods, 1 mile (1.5 km) to the east of Nupend. Like Nupend, these too surmount ridges of Silurian limestone. Formerly mixed oak and ash woods with an understory of hazel, several old oaks still survive in Paget's Wood, but those in the Lea Wood reserve have been felled. The latter is now being coppiced, with stools beneath a canopy of spindly ash and birch. Both woods exhibit a varied, luxuriant understorey that includes field maple, hazel, yew, wild service, spindle, wild rose, bramble and masses of honeysuckle. Wild cherry also flourishes in the south-eastern corner of Paget's Wood, while flowers include wild daffodils in early spring and several orchids in summer. Along the tiny stream is a belt of mature alders.

## Species

Few rare birds can be expected and at times the woods seem almost lifeless. Yet there is a wide range of woodland birds to be enjoyed amidst some superb scenery.

On a mild winter's day, the resident Wrens, Dunnocks, Robins, thrushes, tits and finches utter their first, faltering songs. Overhead the aerial displays of Sparrowhawks sometimes end in a skirmish with a passing Buzzard or Raven that has shown the temerity to drift into their territory.

April brings new sights and sounds every day. First on the scene are Chiffchaffs and Willow Warblers, searching the still-bare branches for food. By mid-month Blackcaps can be heard from deep within the shrub layer, while Cuckoos call across the valley. On the outskirts of the woods Tree Pipits, Garden Warblers, Lesser Whitethroats and Whitethroats add their songs to those of the resident Chaffinches and Bullfinches. Open areas such as Broadmoor Common and Common Hill are good for these species, with Whitethroats, Linnets and Yellowhammers particularly fond of gorse.

Of the scarcer residents, all three woodpeckers can be seen, or more likely heard, though the Lesser Spotted can be hard to find among the canopy. Other hole-nesting residents include a few pairs of Willow Tit in damp places, and both Stock Dove and Nuthatch among the more mature, deciduous timber. Notable among the scarcer migrants are Redstart and Wood Warbler. Redstarts, too, nest in holes, but they prefer more open situations where the male has space for his display flights and prominent perches from which to sing and flycatch. The shivering trills and plaintive pipings of Wood Warblers usually emanate from beneath a closed canopy with little or no undergrowth. Sometimes a passage Pied Flycatcher is seen.

Here and there in commons, large clearings or young forestry, a pair of Grasshopper Warblers nest, though numbers are currently very low. Evenings are often the best time to hear them singing. Indeed, the twilight can be very productive, with Woodcock beginning to rode above the trees and Tawny Owls calling as darkness falls. The real prize, though, is to hear the unmistakably rich, flutey song of a Nightingale, which here is right on the very edge of its breeding range. One or two are heard most years, though not necessarily in the same area. They prefer sheltered spots where the canopy is sufficiently open to permit a luxuriant shrub layer. Coppiced woodland or commons with dense thickets, such as Broadmoor Common, are often favoured.

In the plantations, Coal Tits and Goldcrests are common even in quite dense stands of conifers, while purring Turtle Doves can be heard in conifers and broad-leaves alike. Twittering parties of Redpoll nest high in mature conifers, but feed in the bushy growth around clearings, rides and the woodland edge.

By high summer there is little visible activity, although some early nesters may begin to wander. In irruption years the explosive 'chip' flight calls of Crossbills may be heard from July onwards as small parties feed upside down beneath pine or larch cones. As many as 40 have been seen in Haugh Wood in May, so breeding is a possibility. As summer fades so tits and warblers wander through the woods in search of food. Soon the summer visitors have gone and the resident birds turn increasingly to the forest fruits for their food.

As autumn leaves turn to gold, red and brown, so the winter visitors begin to arrive. October sees migrant Mistle Thrushes, Blackbirds, Fieldfares and Redwings join their resident cousins in plundering yew berries. Lots of finches and sometimes a wintering Blackcap feed with them on yew or honeysuckle. As the grip of winter tightens, so tits and finches rummage through the leaf litter in search of mast beneath the lofty beeches. Great Tits and Chaffinches are commonest, but a Marsh Tit or a few Brambling may join them. Small flocks of Redpoll and variable numbers of Siskin wander restlessly through birchwoods and along streamside alders, sometimes in company with a charm of Goldfinches, or a solitary Treecreeper or Lesser Spotted Woodpecker. On the hillsides, Woodcock lie camouflaged amongst dead, frost-etched fronds of bracken. Before long the days lengthen again and the first, faltering songs become stronger and more assertive. By March, Siskin are leaving their winter haunts to form pre-migration gatherings in the conifer plantations. Most then leave during April, but a few remain into May and a pair or two occasionally breed. With luck, even a shy Hawfinch might be glimpsed in spring.

## Timing

Early morning in spring or mid-summer is the best time to visit as birds are then at their most active. For resident species, early spring, before the trees are in leaf, is best, but for migrants a May visit is recommended. Although Nightingales do sing during the day, the best time to listen for them is late on a warm evening in May or early June. Timing matters less in winter as fewer people are about and birds are generally active throughout the short daylight hours. As with most woodlands, little is seen in very wet or windy weather.

## Access

Leaving Hereford eastwards on the B4224 Mitcheldean road, continue for 4 miles (6.5 km) into Mordiford and then take the second turning left in the village towards Woolhope. Continue up the hill into Haugh Wood and park in the FC car park on the left. From here there are two forest trails that can be followed, one on either side of the road. For Broadmoor Common continue along the same road until it comes onto the common immediately after leaving the wood. Turn right into a narrow lane towards Fownhope, park on the verge and explore from here.

For the two HRNT reserves continue down the hill and then turn right towards Fownhope. For Lea and Paget's Woods turn left at the next junction and continue along a narrow lane for about ⅓ mile (0.5 km). As the road begins to descend the hill there is a gate on the left next to a derelict stone building. This gives access to the reserve, which is restricted to HRNT members only. Parking is extremely difficult, so take great care.

For Nupend Wood carry straight on at the previous junction for a little over 1 mile (2 km) and then park on the verge where the Wye Valley Path crosses the road. Access to the reserve, which again is restricted to HRNT members only, is through the gate on the right. The Wye Valley Path gives ample opportunity to see the characteristic birds of the area in some magnificent scenery and visitors will do as well walking this as visiting the reserves.

## Calendar

*Resident:* Sparrowhawk, Woodcock, Tawny Owl and most woodland birds including Stock Dove, all three woodpeckers, Marsh and Willow Tits, Nuthatch, Treecreeper, Redpoll, Linnet and Yellowhammer.

*April–June:* Turtle Dove, Cuckoo, Tree Pipit, Nightingale (scarce), Redstart, scrub and woodland warblers including Grasshopper (rare) and Wood Warbler, Spotted Flycatcher, Pied Flycatcher (scarce), Siskin (mostly April), Crossbill (scarce) and Hawfinch (rare).

*July–September:* Mostly quiet in July, but Crossbills may irrupt. Tits and warblers rove through woods.

*October–March:* Buzzard, thrushes (including Fieldfare and Redwing), Blackcap (rare), Raven and finches including Brambling (scarce), Siskin (increasingly in conifers after February) and Crossbill.

# HEREFORD LUGG & WYE

Map 30
OS map 149

## Habitat

To the east of Hereford the River Lugg flows through a wide, open flood plain of unimproved pastures. South of the city the Wye too begins to describe broad meanders through its flood plain, which becomes progressively narrower and gorge like as the river flows southwards. The confluence of the two rivers is at Mordiford, some 4 miles (6.5 km) downstream of the city centre. Riverside pastures dominate this open countryside, where hedges are few and trees scarce apart from the bankside willows and alders. As a rule most birds are to be found along the Wye, particularly at Hampton Bishop and Sink Green. During times of flood, however, the Lugg Valley around Tidnor may hold the largest concentrations.

## Species

This is not a consistently good area for birds, but it certainly repays regular watching. Winter wildfowl and passage waders are the main interest, with the former most numerous during periods of flood.

Diving birds are few and irregular. The Wye usually holds up to ten Cormorants and in late autumn three or four Little Grebes and perhaps a Great Crested Grebe. By mid-winter one or two Pochard could have arrived, while small numbers of Goldeneye and Goosander are present at times between January and March.

Dabbling duck are much more variable. Of the common species, a few Mallard are normally present and up to 30 Mute Swans may congregate in autumn or spring. Teal could be present at any time and might even breed. Up to 50 are usual during January and February, but flooding may bring as many as 250. It also attracts a similar number of Mallard, a dozen or more Mute Swans, and maybe a sizeable herd of Bewick's Swans, one or two Whooper Swans or a Pintail. Up to 50 Wigeon sometimes gather between January and March to graze on partially flooded meadows. Shelduck regularly pass along the rivers during their spring or autumn migrations and Shoveler make occasional, brief appearances. Gadwall have been recorded too, but are more likely to be seen downriver in the vicinity of Fawley.

Waders fall into two categories, namely those migrating through in spring and autumn and those visiting the flood meadows in winter. Of the migrants, Curlew visit the Lugg Meadows during March and many Common Sandpipers pass up the Wye during April. Small numbers of Oystercatcher, Dunlin and Redshank may move with the latter, but they are erratic and rarely stay long. Return passage is generally more leisurely and varied, especially when the river is low. By July both Green and Common Sandpipers are moving back downstream and one or two can then be expected until October. A Green Sandpiper might even stay through a mild winter. Two or three Redshank, possibly a local family, may also appear in July, to be followed in late autumn perhaps by another that might loiter into the New Year. August is typically the best month for waders, with very small numbers of Dunlin, around a dozen Common Sandpipers and up to ten Greenshank in a good year.

In winter the riverside meadows regularly hold well over 1,000 Lapwing, with three times that number on occasions. With them may be a couple of dozen Golden Plover. A few Snipe feed along the muddy river banks, but many more probe the meadows for food as floodwaters subside, with wisps of 20 by no means unusual. Exceptionally one or two Dunlin, or even a Jack Snipe might be seen. As well as waders, up to 1,000 Black-headed Gulls and 50 Common Gulls feed in the meadows each winter. Thrushes, too, are regular visitors, with large flocks of 1,000 or more Fieldfare and Redwing in late autumn and winter. With them small parties of migrating Skylarks and Meadow Pipits move westwards in autumn and return again in spring. Some may even winter among the damp pastures and several pairs of Skylarks breed.

Kingfishers nest in the river banks and can be seen passing upstream and downstream at any time. Good numbers of wagtails are often present too. Pied Wagtails are always about, but are most numerous in late autumn, when as many as 100 may be in the area. Small numbers of Grey Wagtails also pass through at this time, but they are scarcer in spring as they make for their upland breeding streams. Conversely, the summer-visiting Yellow Wagtail is more evident in spring, when the brightly-coloured males attract attention. Most common summer migrants pass through in small numbers. Sand Martins, Swallows and House Martins hawk insects across the flower-rich meadows, Whinchats and Wheatears flit across the ground, and the occasional Redstart is glimpsed in riverside willows. Perhaps a Stonechat might visit in late autumn. From late summer onwards, small parties of Linnets and Goldfinches come to feed, the latter especially on thistle heads. As the winter sets in, these parties decline in number, but 50 or more Chaffinches, a few Tree Sparrows and Yellowhammers, and less often one or two Bramblings or Reed Buntings add variety. Kestrel apart, this is not an outstanding area for predators, though a Barn Owl may still be seen searching for prey.

Although principally an area for open ground species, scrub and woodland birds also occur, some of them quite regularly. The alders at Hampton Bishop hold a few Redpoll and good numbers of Siskin in winter, and Lesser Spotted Woodpeckers are seen quite often around the same village. Passage warblers use the limited cover available, with Lesser Whitethroats regular summer visitors. Rarities also occur from time-to-time and in recent years these have included Hoopoe, Great Grey Shrike and, most remarkably, a Killdeer on the Wye in January.

## Timing

Being so close to Hereford, this is a well-used area, but weather has a greater influence on birds than disturbance. Spring passage is stronger in poor weather, when birds such as hirundines are forced to follow the river valleys and to fly and feed low. For waders, low water levels expose more mud and shingle spits in the channels and few birds can be expected when the river is bank high after very heavy rainfall. Several days of heavy winter rainfall or a rapid snow thaw are needed to bring the flooding necessary to attract large numbers of wildfowl. Receding floodwaters then bring in waders, thrushes and maybe other open ground feeders. Severe frost will quickly force birds to vacate the meadows which become too hard to probe for food.

## Access

There are several access points from the four main roads that cross the area. The south bank of the Wye can be reached from the B4399 just south of Sink Green, from where there is a footpath alongside the river in both directions. Likewise the north bank can be reached from footpaths that leave the B4224 Hereford to Mitcheldean road immediately west of Mordiford Bridge and just west of Hampton Bishop.

The Lugg flood meadows are best explored from the A4103 Hereford to Worcester road (parking is difficult), from which footpaths leave to north and south at Lugg Bridge, or from the footpaths around Hampton Bishop. During times of flood it may be necessary to explore from surrounding roads, when the narrow lane that runs to the north-east of the river between Lugwardine Bridge and Mordiford can be useful.

## Calendar

*Resident:* Cormorant, Mute Swan, Teal, Mallard, Kestrel, Kingfisher, Lesser Spotted Woodpecker and Pied Wagtail.

*March–June:* Shelduck, Oystercatcher (scarce), Dunlin, Curlew (March), Redshank, Common Sandpiper, Skylark and Meadow Pipit. From mid-April hirundines, Yellow Wagtail, Lesser Whitethroat and maybe Redstart, Whinchat and Wheatear.

*July–September:* Shelduck, Dunlin, Redshank, Greenshank, Green Sandpiper, Common Sandpiper, hirundines, Yellow Wagtail, Redstart, Whinchat, Wheatear and flocks of Goldfinch and Linnet in September.

*October–February:* Little and Great Crested Grebes, Bewick's Swan (erratic), Whooper Swan (rare), Wigeon, Gadwall (rare), Pintail (rare), Shoveler (erratic), Pochard, Goldeneye, Goosander, Golden Plover, Lapwing, Jack Snipe (rare), Snipe, Redshank (scarce), Green Sandpiper (scarce), Black-headed Gull, Common Gull, Barn Owl (rare), Skylark and Meadow Pipit (mostly October-November), Grey Wagtail, Stonechat (scarce), Fieldfare, Redwing and flocks of finches and buntings including Siskin, Redpoll and maybe Brambling and Reed Bunting.

## Habitat

Throughout its entire 156 mile (250 km) course, the Wye must rank as one of Britain's most beautiful rivers. Yet even amidst all its splendour no stretch can compare with the spectacular gorge that begins just below Ross-on-Wye.

Falling sea levels caused the river to impose its sweeping meanders onto the bedrock and cut deeply into the Old Red Sandstone and Carboniferous limestone. The result is a magnificent gorge which can be seen at its best at Symonds Yat. There are outstanding geological exposures including, on the Herefordshire side, the Seven Sisters — a series of massive limestone buttresses with a rich and rare flora. Semi-natural broad-leaved woodland clings precariously to the precipitous slopes. Save for coppicing and the introduction of a few alien species, these woods have been little modified by man and are ecologically among the most important in Britain. They are very varied, with ash, beech, wych elm, small-leaved lime and yew on the more acidic summits; ash, wych elm and hazel on the slopes; alder along the damp, fertile valley flood plain; and oak and beech on the Old Red Sandstone. Wild cherry is also widespread and both the shrub and field layers are rich and varied.

Above the gorge, the huge limestone dome of the Great Doward supports an interesting mixture of woodland, scrub and small meadows. Despite the presence of wild servicetree, the woods are thought to be a secondary recolonisation of previously cleared ground. The Herefordshire and Radnorshire Nature Trust (HRNT) has two small reserves here, which embrace a range of woodland. At Woodside high forest of oak and beech stands above an understorey that contains many berry-bearing species such as blackthorn, hawthorn, holly and yew. Parts are almost pure beechwood, with a typically sparse ground cover, but others have been recently coppiced. In contrast, a long-neglected tangle of coppiced oak and hazel is now reverting to scrub woodland at Leeping Stocks, where some fine beeches mark the old field boundaries. Ground cover varies according to the soil and amount of shade cast by the canopy, with acid-loving species such as birch, bracken and common cow-wheat sometimes standing close to calcicoles like sanicle, marjoram, wood spurge and wood melick.

## Species

The Wye Gorge is best known for its well-publicised Peregrine eyrie and is well worth visiting for this alone, although the surrounding woods are also of interest.

The Peregrine eyrie is on the Gloucestershire side of the river and is best watched from the RSBP's observation post on Symonds Yat Rock. Although the adults are sedentary, observation is easiest during the breeding season. Then the pair are very noisy, often announcing their presence with a shrill, repetitive 'kak-kak-kak' long before you actually catch sight of them. Their behaviour can only be described as spectacular, with steep climbs, breath-taking plunges, acrobatic rolls, talon gripping and food passes. Most sightings are of the smaller male as he

*The spectacular display of Peregrines can be watched above the Wye Gorge*

sets off along the valley with stiff, winnowing wing-beats, before stooping headlong on folded wings to strike his quarry with awesome speed and accuracy. Outside the breeding season you might be lucky enough to see one of the pair soaring high overhead or hanging on an updraught with characteristic anchor shape.

Peregrines apart, there is a good range of scrub and woodland birds. Jackdaws nest in crevices amongst the limestone crags and noisy parties are constantly wheeling around the cliffs. The scrub and woods along the gorge and on the Doward are full of the usual Wrens, Dunnocks, Robins, thrushes, tits and finches. All three woodpeckers, Nuthatch and Treecreeper are resident, but keep mostly to the more mature woodland. In spring the drumming of woodpeckers echoes across the gorge. Stock Doves and Tawny Owls nest in the older trees, while Woodcock emerge from their day-time roosts to feed at dusk. Summer visitors include Tree Pipits and Redstarts around the woodland edge and most of the scrub and sylvan warblers, including one or two pairs of Wood Warbler under the beeches. Willow Warblers are especially widespread and, along with Chaffinch, are the commonest bird in the ashwoods. On the whole, Leeping Stocks is best for scrub species like Whitethroat, Garden Warbler, Linnet and Yellowhammer, and Woodside for arboreal ones like Chiffchaff, Blackcap and Jay. In autumn, long after the summer visitors have left, Redwing and Fieldfare arrive to feed on the berries, but numbers are not very large. A few Mistle Thrushes are usually attracted by the yew berries.

The river is less interesting than it is above Ross-on-Wye, but Cormorants and Grey Herons regularly fly past, often at great height. Even higher above the valley, a Raven might make purposeful progress on slow, powerful wing-beats, its wedge-shaped tail clearly discernible against the sky. Many birds follow the Wye on their migrations and Swallows and House Martins can sometimes be seen in good numbers taking insects from above the water. Recent sightings of both Hoopoe and Black Kite should stimulate anyone thinking of exploring this little-known district to do so.

## Timing

Symonds Yat gets very congested, so avoid fine summer weekends and Bank Holidays if at all possible. The Peregrines, though, can be watched at any time, but are most active when feeding young in late May or June. For woodland birds, early mornings are the best time of day and May the best month. Little will be seen in wet or windy weather.

## Access

Symonds Yat is best approached from the A40 Ross to Monmouth road: 3 miles (5 km) beyond the western end of the Ross by-pass turn southwards towards Goodrich. In the village turn right onto the B4229 and in another ¾ mile (1.2 km) turn left onto a minor road. Cross over the river and then in 1 mile (1.6 km) fork left up a narrow steep hill to Yat Rock. There is plenty of car parking at the summit on both sides of the road.

For the Doward reserves continue south-westwards on the A40 to Whitchurch. Leave the main road here and take the old road to Crockers Ash. Then turn southwards into a narrow lane past the Doward Hotel. Continue along this lane, round both right-hand and left-hand hairpin bends. Shortly after the second bend take the unsurfaced track straight ahead and the Leeping Stocks reserve is 160 yds (150 m) on the left. Car parking is extremely difficult and access to the reserve is strictly for HRNT members only.

For Woodside continue on the narrow lane for another 200 yds (180 m) beyond the track to Leeping Stocks. Park where the lane turns sharp left and follow the track straight ahead up the hill. At the fork turn right and the reserve is on the right. Again access is strictly for HRNT members only.

## Calendar

*Resident:* Cormorant, Grey Heron, Peregrine, Woodcock, Stock Dove, Tawny Owl, all three woodpeckers, Nuthatch, Treecreeper, Jackdaw, Raven and common woodland passerines.

*April–June:* Swallow, House Martin, Tree Pipit, Redstart and scrub and sylvan warblers including Wood Warbler. Maybe a rare migrant.

*July–September:* Quiet in July. Summer visitors leave in August and September, when Swallows and House Martins pass along the river. Maybe a rarity.

*October–March:* Parties of thrushes (including Fieldfare and Redwing), tits and finches.

## Habitat

The Black Mountains protrude into west Herefordshire, where, at 2,306 ft (703 m), they form the highest part of the region. The county boundary is marked by the Offa's Dyke Path, which wends its way across a high, precipitous ridge. To the west is the Brecon Beacons National Park, to the east the glaciated Olchon Valley. This remote, beautiful valley is penetrated only by a steep and narrow lane which loops back on itself to return on the opposite hillside.

Along the valley the tiny, babbling Olchon Brook is flanked by a patchwork of small, fertile meadows. These continue up the gentle lower slopes, but gradually give way to banks of bracken as the gradient steepens. Above the bracken are scree slopes surmounted by sheer cliffs and crags of exposed sandstone. These open out onto a heavily-dissected, windswept plateau of heather moor in which crowberry is locally abundant.

Woodland is not extensive, but there are some open ashwoods along the higher slopes and tiny sessile oakwoods lower down. Moreover there is a wealth of free-standing and hedgerow timber. Stunted hawthorns are characteristic of the moorland fringe, particularly high in the cwm at the head of the valley

## Species

The primary interest is upland breeding species. On the high tops, where winter can be harsh, truly resident birds are few. But there are exceptions, notably Red Grouse and corvids. The Black Mountains hold fair stocks of Red Grouse and birds are quite often noted on the plateau above the Olchon. Autumn is a good time to hear their noisy cackling as a covey skims across the heather and disappears into dead ground.

*Red Grouse are one of the few resident birds on the windswept Black Mountains*

Overhead a Raven croaks as it moves purposefully across the valley in search of carrion. Ravens nest early, so their tumbling, diving display flight can be seen in winter as well as early spring. Later, in autumn, as many as 20 might roost together or flock around a good food source.

At the approach of spring, Curlews and Meadow Pipits are the first birds to return to their upland breeding grounds. From mid-March onwards, small parties pass up the valley and before long Meadow Pipits are everywhere across the moor. Close behind them, the first Wheatears and Ring Ouzels may arrive while snow still lies in the shadow of sheer cliffs. Two or three pairs of Ring Ouzels usually occupy territories on the high crags and scree slopes from where their shrill, piping song carries far across the valley beneath. Sharing their habitat, a few pairs of Wheatear bob restlessly from boulder to boulder.

Soon, other migrants return to the valley. Willow Warblers are numerous and widespread, Chiffchaffs sing from the tree-tops and Blackcaps warble from bushy undergrowth. From the open woodland high on the valley slopes, Tree Pipits sing and the scratchy notes of Whitethroats descend from patches of scrub. Here and there Whinchats nest, the male scolding any intruders as he flits from perch to perch. Along the brook, Grey Wagtails and Dippers busily feed young, while in the overhanging branches Redstarts and Pied Flycatchers dart after insects. Half-a-dozen pairs of Pied Flycatchers breed in nest-boxes at the head of the valley. Most common birds can also be found somewhere along the valley, including all three woodpeckers.

There is plenty to see and hear in spring, but after mid-summer birds become less obvious. They then slip quietly away in late summer, when a family party of chats, warblers or flycatchers might feed around the same spot for a few hours before moving on again. As the days shorten and the winds sweep storm clouds across the high tops, so hundreds of Redwings and Fieldfares, and occasionally a few dozen Siskin shelter in the valley.

Raptors can always be expected, especially Sparrowhawk, Kestrel and Buzzard. Less often a Merlin dashes across the heather with fast, winnowing wing-beats, then swoops down the valley in pursuit of its prey. The Brecon Beacons have long been a Merlin stronghold, but numbers have fallen with the recent national decline. Peregrines, though, are enjoying a national resurgence and one of these majestic falcons might soar above the high crags at any time. Exceptionally, spring or autumn bring other wandering raptors like a Short-eared Owl.

## Timing

Weather is the most critical factor. Although the first summer visitors can arrive as early as March, in a cold, late spring they may be delayed until April or even May. Early mornings are the best time of day, before too many people are in the valley or on the moor. If bad weather is forecast it is advisable to think twice about venturing onto the summit, where walking is rough and it is easy to get lost in the mist. Nothing much will be seen in wet, misty weather anyway. On windy days birds tend to seek the shelter of the valley.

## Access

The best approach is from the A465 Hereford to Abergavenny road. Turn north-westwards into a narrow lane at Pandy, which is 6 miles (10 km) north of Abergavenny. Pass under the railway, over the river and then turn right at the T junction. Continue straight ahead for 3 miles (4.5 km) to Clodock. Turn left in the village, and follow the road to Longtown, forking left again in ⅔ mile (1 km). Go through Longtown village and in ¾ mile (1.2 km) turn left again down a steep hill and across the brook. Follow the lane round to the right and beneath the towering outcrops of the Black and Red Darens, pausing to explore from the picnic area on the left. Continue past some small woods and at the end turn left to follow the lane into the cwm, pausing again to explore by the bridge across the brook. Leave the valley on the opposite side and take a track to the left which climbs steeply to another picnic area on the edge of Black Hill. With commanding views, this is a good spot to watch for raptors. There are several footpaths up to the summit plateau, but the easiest ascent is to begin at the bridge in the cwm and follow the track that leaves the lane a few yards to the east.

Alternatively, leave Hay-on-Wye on the B4350 towards Brecon and take the first left after the B4348. Continue climbing up a twisty, narrow lane for 2½ miles (4 km), then take the right fork up a steep hill onto the unfenced moor. In just over 1 mile (2 km) the Offa's Dyke Path crosses the road and ¼ mile (400 m) further on is a car park in an old quarry, just before the road forks. Park here, follow the road back and take the Offa's Dyke Path south-eastwards. This crosses the moor and in just over 2 miles (4 km) another path strikes off left and down into the Olchon Valley.

## Calendar

*Resident:* Sparrowhawk, Buzzard, Kestrel, Red Grouse, all three woodpeckers, Grey Wagtail, Dipper and corvids including Raven.

*March–June:* Curlew, Meadow Pipit, Wheatear and Ring Ouzel. From mid-April Tree Pipit, Redstart, Whinchat, warblers and Pied Flycatcher. Perhaps passing raptors.

*July–September:* Generally quiet, but family parties or mixed flocks of departing migrants in August and September. Perhaps passing raptors.

*October–February:* Winter thrushes and Siskin. Perhaps a passing raptor.

## Habitat

In its middle reaches around Hay-on-Wye, the Wye leaves Wales to enter Herefordshire and England. To a point some 6 miles (10 km) upstream of Hay, the river flows south-eastwards across mineral-deficient bedrocks within a narrow valley. As it bursts forth onto the nutrient-rich Old Red Sandstones, it is abruptly deflected north-eastwards by the towering heights of the Black Mountains. For the next 10 miles (16 km) it sweeps in broad meanders across a mile-wide flood plain, before turning eastwards towards Hereford. During this swift transition, life in the river increases sharply, making it much more attractive to birds.

The river is broad, but fairly shallow, with a gravelly bed. At times there are fast currents across shallow runs, at others a sluggish flow through deeper pools. The meanders have deep, undercut sandstone banks on their outsides, and exposed gravelly spits on their insides. Behind these spits are natural levees and flood meadows backed by some beautiful, rolling, wooded countryside. Within the flood plain there are a few large willows and alders.

## Species

Herefordshire has few pools or lakes, so wildfowl and waders make great use of suitable stretches of river such as this. Numbers are seldom large and birds cannot be guaranteed, as much depends on the state of the river and the flood meadows. Nevertheless, regular watching yields an interesting list of species.

Mute Swans are a feature of this stretch of river. Only a pair or two breed, but from late summer through to spring small herds of a dozen or so gather at various points along the valley, sometimes merging into concentrations of 50 or more. Small gaggles of Canada Geese also form in late summer, perhaps building to a November peak of 200, then declining to just 30 or 40 through the winter.

These swans and geese sometimes draw other wildfowl into the valley. A few Whooper Swans are seen most winters, often just across the Welsh border around Glasbury. Bewick's Swans are less regular, but from time-to-time a small party, often a family, stays for just a couple of days. White-fronted Geese occasionally do likewise, while very rarely a Pink-footed or other goose passes by.

Dabbling duck numbers vary with the weather and the degree of flooding. Mallard are most consistent, with several pairs breeding and some 200 present from September through to February, but Teal and Wigeon fluctuate according to conditions. A few Teal can often be seen from September onwards, but parties above a dozen are unusual before January or February, when flocks of 50 or more come to the flood meadows. By March most have left again. Wigeon seldom appear much before November, when a couple of dozen may settle in for the winter. If the weather stays mild, no more may arrive, but in flood conditions or severe cold flocks of 100 or more whistle evocatively across the meadows during February and March.

Diving birds are confined to the river and pools and do not depend on floodng. Up to ten Cormorants and about half as many Little Grebes are

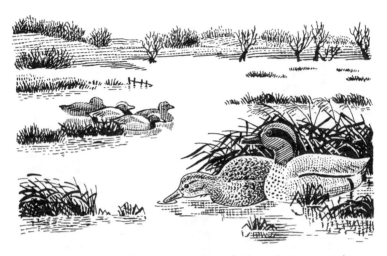

*Flocks of dabbling duck, such as Wigeon (left) and Teal, come to the flood meadows of the Wye in winter*

regularly present from September through to March. Small parties of Tufted Duck also occur, with possibly 40 in a flock, but Pochard are scarce, with no more than six or so in mid-winter. Great Crested Grebes are uncommon too, though a pair might appear in spring. Goldeneye and Goosander, though, are regular visitors. Five or six Goldeneye arrive in November and stay until March, being joined perhaps by another dozen in severe weather. The Wye is very good for Goosanders. One or two early birds begin to move through in late August or September, when exceptionally parties of 30 might be seen. Few stay around, however, and it is October or November before the first wintering birds arrive, with some six or so regularly along this stretch of river. Then in March and April parties up to 20 strong again occur as birds begin to return to their breeding grounds. As with Goldeneye, numbers often increase in hard weather.

Grey Herons from the small heronry near Clifford frequently fly up and down the river or stand patiently along the riverbanks waiting for fish. Kingfishers, too, are ever present and might be encountered at any time. Outside the breeding season, small parties of Black-headed Gulls frequent the flood meadows, their numbers highest in winter when they are joined by a few Common Gulls and one or two Lesser Black-backed, Herring and Great Black-backed Gulls.

Lapwings begin to return from their breeding grounds in July and during late summer and early autumn up to 600 feed in the area. Thereafter numbers vary, holding up in some years but declining in others. With the Lapwings are likely to be a few Golden Plover, perhaps even 400 on passage in November, although the wintering flock seldom exceeds 100 and in a poor year is less than ten. Up to a dozen Snipe arrive in August and they too winter in the valley, their numbers increasing two or threefold when floodwater is standing on the meadows. Very rarely a Jack Snipe is found in November or December. Some 30 or more Curlew over-winter upstream of Hay, while the Castleton area holds an occasional Woodcock and up to four wintering Green Sandpipers, which frequent the small pools and ox-bow lakes.

March and April are the main months for passage waders, with solitary Dunlin and Redshank regularly recorded and parties of six to twelve Common Sandpipers working their way upstream along the river banks in April. By late July, return passage has begun. First to show are usually a couple of Green Sandpipers and up to half-a-dozen Common Sandpipers, followed in August by the main passage of Common Sandpipers and Greenshanks. Five or six of the latter are normal, but 20 to 30 may be present in a good year. Accompanying these regular visitors may be a single Redshank or Ruff, but Dunlin are surprisingly rare in autumn. One or two Arctic and Common Terns also migrate along the river, with more in autumn than spring.

Many migrants follow the sheltered valley, especially during a cold spring when numbers can be good. Sand Martins begin to return to their riverbank nest-holes in late March and the first Wheatears arrive about the same time on their way to the hills. Shortly afterwards a few brilliant Yellow Wagtails appear in the lush meadows, one or two Redstarts stake out their territories and the occasional pair of Sedge Warblers takes up residence. By late summer and autumn, families of the same species leisurely make their way back down river. With them may be a family of Whinchats or Spotted Flycatchers, a stray Whitethroat or Lesser Whitethroat, or a solitary Pied Flycatcher.

As the summer visitors leave, so resident songbirds like Linnet and Goldfinch flock together. Later, as the winter floods subside, small flocks of larks, pipits, finches, buntings, and 100 or so Fieldfare and Redwing gather to feed. Parties of Siskin and Long-tailed Tits sometimes work their way through riverside alders and both Pied and Grey Wagtails feed along the banks. Ravens, Stock Doves and all three woodpeckers also frequent the area.

The abundance of wildlife draws raptors to the valley. Kestrel, Sparrowhawk and Little Owl are regularly seen, Buzzards glide overhead and in the evening twilight a Barn Owl sometimes floats across the flood plain. Less often a passing Merlin or Peregrine pauses to harass the wintering wildfowl and waders. Rarities are few, but spring may bring a Water Pipit or White Wagtail, while a vagrant such as Kittiwake could always appear after an autumn or winter gale.

## Timing

Apart from anglers, there is little human disturbance. In winter flooding is critical. Normally it does not occur until the end of the year, and conditions are usually best sometime between January and March. Hard weather may bring an influx of diving duck from frozen pools and lakes. Spring passage is often best in cool, wet weather, when birds fly low and follow the valley. Periods after gales promise to be interesting, whether in spring or autumn.

## Access

The 14 miles (22 km) between Hay-on-Wye and Bredwardine are the most interesting, with the precise whereabouts of birds depending very much on the time of year and the extent of flooding. North of the valley is the A438 Hereford to Brecon road and south is the B4352 Hereford to Hay road. A few river meadows can be viewed from these roads, but the best access points are:

*Hay-on-Wye* Leave the B4351 Clyro road immediately north of the river bridge in Hay and take the Offa's Dyke Path to the right, following this along the Powys bank for 2 miles (3 km) through to the A438.

*Castleton* Take the B4350 (Leominster road) north-eastwards out of Hay as far as Clifford. At the far end of the village turn right into a narrow lane and then in 300 yds (0.3 km) turn left. Continue for a little over 1 mile (2 km) until the lane turns sharp right up a hill. Park near this turn (space is restricted) and follow the public footpaths either to the north as far as the toll bridge on the B4350 — just under 1 mile (1.5 km) — or to the east as far as Clock Mills on the B4352, 2 miles (3 km).

*Winforton* A public footpath leads down to the river from the A438 at the western end of the village.

*Bredwardine* Turn eastwards at the crossroads in the village, cross over the river bridge and take the left turn towards Letton. Alternatively, turn right just before the river bridge, park by the church and follow the public footpath to the right. This skirts the edge of a wood and eventually rejoins the B4352 in a little over 1 mile (1.8 km), opposite Moccas Deer Park. There is no public access to the Deer Park, but it is always worth checking from the road.

## Calendar

*Resident:* Grey Heron, Mute Swan, Canada Goose, Mallard, Sparrow-hawk, Buzzard, Kestrel, Stock Dove, Barn Owl, Little Owl, Kingfisher, all three woodpeckers, Grey and Pied Wagtails, Long-tailed Tit, Raven and other common passerines.

*March–June:* Winter visitors leave in March and April, when passage Great Crested Grebe (scarce), Goosander, Dunlin and Redshank. From late-March, Common Sandpiper, Sand Martin, Yellow Wagtail, Redstart, Wheatear and warblers including Sedge Warbler. Perhaps a rarity.

*July–September:* Summer visitors depart. Wildfowl including early Goosander, Lapwing, Dunlin (scarce), Ruff (scarce), Snipe, Redshank, Greenshank, Green Sandpiper, Common Sandpiper, Black-headed Gull, terns (scarce), Goldfinch and Linnet.

*October–February:* Little Grebe, Cormorant, Bewick's Swan (rare), Whooper Swan (scarce), Wigeon, Teal, Pochard (scarce), Tufted Duck, Goldeneye, Goosander, Golden Plover, Lapwing, Jack Snipe (rare), Snipe, Woodcock (scarce), Curlew, Green Sandpiper, gulls and flocks of passerines including Fieldfare, Redwing and Siskin. Maybe a passing raptor.

# EYWOOD POOL

**Map 34**
**OS map 148**

### Habitat

Eywood Pool although modified by man, is believed to be of glacial origin. Standing water is scarce in Herefordshire, so this delightfully secluded 15 acres (6 ha) pool, set in private parkland, is of special importance. There is a small island and a narrow belt of perimeter woodland that includes many oaks and some alder, ash and birch. Owned by the Herefordshire and Radnorshire Nature Trust (HRNT), the reserve has recently been replanted with hornbeam and beech. The luxuriant shrub layer attracts many breeding birds, especially warblers.

### Species

Wildfowl are the main interest. Numbers are never high, but regular watching has revealed a good variety.

One or two pairs of Great Crested Grebe, Mute Swan, Canada Goose, Mallard and Coot regularly breed. By early autumn some 150 Mallard and 50 Tufted Duck may be present, with smaller numbers remaining through the winter alongside up to 30 Pochard and 50 Coot. A couple of dozen Teal normally arrive around September and if the pool stays open they may remain until the following spring. If it freezes, though, they quickly leave, although a few may return again in February or March.

Other species are less regular, but one or two Shoveler and Goldeneye pass through, usually staying just for a day or two in autumn and returning in March. Recently a few Ruddy Duck have been noted, particularly in late spring and summer, and future breeding seems probable. Erratic visitors have included Cormorant, a party of eight Barnacle Geese (origin unknown), Garganey, Pintail and Goosander, while sea-duck appear from time-to-time. Wildfowl apart, both Grey Herons and Kingfishers often come to seek their next meal. There is little suitable habitat for waders, but a muddy edge in late summer might hold a Green or Common Sandpiper.

Around the pool, Little Owl, Great and Lesser Spotted Woodpeckers, and Nuthatch are resident along with a host of common species, while Buzzards frequently circle and soar overhead. Spring passage is normally good, bringing Turtle Dove, Redstart, Blackcap, Wood Warbler and Pied Flycatcher along with commoner migrants. Winter is quieter, but a Woodcock might be flushed from the wood or watched emerging to feed at dusk. In recent years passage Dotterel have twice appeared on Bradnor Hill, 2 miles (3 km) to the west, in May and August and it is possible that 'trips' of this species regularly stop-over here.

### Timing

There is little disturbance, but early morning is always the best time to hear and see songbirds in spring and summer. Duck are most numerous in late summer and autumn, but desert the pool altogether when it freezes over. Because it is quiet and secluded, many wildfowl use the pool solely for roosting, so late evening may bring some fresh arrivals from elsewhere.

## Access

Eywood Pool is owned by the HRNT and the surrounding land is all private. The water is small and anyone approaching the shoreline is liable to disturb the birds. For this reason access is strictly limited to HRNT members only, who may use the hide provided.

Approach from B4355 Kington to Presteigne road, turning westwards between Titley Court and the Stag Inn into the Eywood Park entrance. Drive past the pool on your left, take the first turning left, cross a cattle grid and turn left again after 50 yds (50 m). Go through two field gates and drive on to park in the space provided on the right, just before the cottage. The hide is on the bank immediately below.

## Calendar

*Resident:* Great Crested Grebe, Mute Swan, Canada Goose, Mallard, Buzzard, Coot, Little Owl, Kingfisher and most common woodland birds including Great and Lesser Spotted Woodpeckers and Nuthatch.

*April–June:* Ruddy Duck, Turtle Dove, Redstart, woodland warblers including Wood Warbler, and Pied Flycatcher. Perhaps a passage Dotterel on nearby Bradnor Hill.

*July–August:* Grey Heron, perhaps a passage Green or Common Sandpiper, or maybe a Dotterel on nearby Bradnor Hill.

*September–March:* Grey Heron, Teal, Shoveler (sporadic), Pochard, Goldeneye (sporadic), possible Goosander or scarce wildfowl, and Woodcock.

The southern areas of Shropshire are poorly recorded ornithologically, perhaps because they are mostly remote from centres of population. Yet this fact should surely add to the appeal of this lovely district of low wooded hills and valleys, with such delightful rivers as the Teme and its tributaries Clun and Onny, and scores of lesser streams.

Ludlow, a beautiful and historic old town, is a good centre for this region, which straddles the boundary between Shropshire and Hereford and Worcester. Short car journeys will take one to the Clee Hills and Wenlock Edge, the many patches of woodland that make up Mortimer Forest, low-lying Wigmore, the commons of Bircher and Leinthall, and dozens of other sites worthy of exploration. The town itself is bordered on the south and west by the River Teme, graced by Kingfishers and Grey Wagtails, and beyond the waters of which lies Whitcliffe Common, famous for its wintering flock of Hawfinches.

## WHITCLIFFE COMMON & RIVER TEME
### Habitat
Although a small area, cramped between the River Teme and the Ludlow to Wigmore road, Whitcliffe offers good year-round birdwatching. The common overlooks the town from high level, and the undisturbed grass is dotted with thorns. The steep slope with rock outcrops, which plunges to the river, is thickly wooded with broad-leaved trees and a few conifers. A gully, aligned with Dinham Bridge, is lined with hornbeams, the seeds of which are avidly sought after by Hawfinches and Bullfinches. The river is lined with alders and the water is rather tamed by weirs situated at both ends of this stretch.

### Species
The river is rarely empty of birds. Grey Wagtails, Pied Wagtails and Dippers feed from the weirs, and the former are sometimes present in numbers at Ludford Bridge. The unmistakable brilliance of a Kingfisher will attract attention as a bird arrows low over the water, yet it will be most inconspicuous when perched motionless in the shadow of riverside vegetation. Moorhens leave the water to feed in the small meadows on the town side bank, and Little Grebes dive in the seclusion of overhanging branches near Dinham Bridge. Small numbers of Mallard are present throughout the year, and the alders are frequented by flocks of Siskins and Redpolls in winter.

The open common is not extensive enough to hold many birds. Even so, Wrens, Dunnocks, Robins, Blackbirds, Willow Warblers, Chaffinches and Yellowhammers are seen in the bushes, Kestrels hunt in quiet periods, whilst Swifts and hirundines hawk the area for insect prey. Many species may be seen from here, such as wildfowl following the Teme, and Grey Herons, Buzzards and Ravens which frequently visit the area.

*At Whitcliffe, Hawfinches can be watched against the backcloth of Ludlow Castle*

It is the woodland that holds the main attractions. Hawfinches in the hornbeams, Siskins and Redpolls in the birches and larches, and even occasional Crossbills, brighten up winter birdwatching. Commoner finches are present throughout the year, along with resident thrushes, which are joined by Fieldfares and Redwings in autumn and winter. Bramblings arrive in some winters. All three woodpeckers occur, as do Nuthatches and Treecreepers, and six species of tits can be found. Of the summer migrants, Whitethroats, Garden Warblers, Blackcaps, Chiffchaffs and numerous Willow Warblers breed, all being well established before the Spotted Flycatchers arrive in early May. Wrens and Dunnocks skulk in the undergrowth whilst Goldcrests seek food in the tops of conifers, especially at the northern tip of the common. Jays, handsome but noisy, are on hand to plunder the nests of some unfortunate victims, and such a proliferation of small birds attracts Sparrowhawks into the area on hunting missions.

## Timing

Whitcliffe Common is a great favourite with both locals and visitors, and must be the most popular dog-exercising spot in Shropshire. The river is also a choice spot with anglers. However, the nature of the woodland, spread along the steep slope between the two, allows all but the very shyest birds to continue their daily routines undisturbed. It is a good area throughout the year, but especially in spring and summer with an abundance of breeding species, and also provides good winter birdwatching with flocks of thrushes, tits and finches. Hawfinches, the speciality of Whitcliffe, are present from November until March, peaking in January and February. The flock often topped 40 birds until the early 1980s, but has been much smaller in recent years.

For completely undisturbed birdwatching, very early morning is essential during the months of long daylight hours, and evenings offer

| Lin | John | Jean | Rich | John |
|-----|------|------|------|------|
| 201 | 86 | 135 | | |
| 359 | 97 | 25 | | |
| | 181 | 98. | | |
| | | 12. | | |
| | | 228 | | |
| | | 228. | | |
| | | 164 | | |
| | | 392 | | |
| | | 782 | | |

a good alternative; in summer, mid-day will produce the least bird activity.

## Access

Walk from the castle in Ludlow to cross the Teme at Dinham Bridge. Alternatively, take the unclassified Ludlow to Wigmore road which leaves the B4361 (formerly the A49(T) before the by-pass was constructed) west immediately south of Ludford Bridge. There is a parking area in the woodland on the left and verge parking space on the edge of the common.

## Calendar

*Resident:* Mallard, Sparrowhawk, Kestrel, Moorhen, Kingfisher, three woodpeckers, Nuthatch, Treecreeper, Grey Wagtail, Pied Wagtail, Dipper, resident woodland songbirds, Goldcrest, six tits, Jay, resident finches, Yellowhammer.

*April–June:* Resident species breeding, followed by Willow Warbler, Whitethroat, Garden Warbler, Blackcap. Spotted Flycatcher from May. Feeding Swift and hirundines.

*July–September:* Mixed flocks of juvenile tits. Finch flocks begin to form.

*October–March:* Little Grebe. Fieldfare and Redwing from mid-October. Winter finches, including Hawfinch, Siskin, Redpoll, Brambling, and occasionally, Crossbill. Chiffchaff from end of March.

## MARY KNOLL VALLEY & MORTIMER FOREST

### Habitat

Mortimer Forest is made up of many separate woodlands of varying size. One large area commences from the western tip of Whitcliffe Common and runs westward for some 4¾ miles (7.6 km) as Bringewood Chase. At the eastern end the forest swells southwards for 2 miles (3.2 km) to the west of Richard's Castle, where it contains the Mary Knoll Valley, Haye Park Wood, and the highest point in the area, High Vinnalls, which reaches 1,235 ft (376 m). Practically the whole area is blanketed with conifers, although a few broad-leaved sections remain, and the Mary Knoll Valley offers several of these. There is a large area of oak near the entrance to the valley at Overton, and the public footpath descending south-east from Mary Knoll House on the Ludlow to Wigmore road passes through an interesting hillside wood of mixed broad-leaves, including oak and birch.

Monoculture on a large scale does not offer exciting birdwatching, but some variety is provided by firebreaks and extraction roads, often lined with weeds, which attract tits and finches. Cleared areas, although usually quickly replanted, make good birdwatching sites for a few years. Adders are reputed to bask along the forest tracks and roads, and fallow deer are numerous. The forest contains a unique long-haired variety of the latter.

## Species

In the vast coniferous areas Goldcrests and Coal Tits are numerous, and along firebreaks and forest roads are joined by other small passerines, including Chiffchaffs and Willow Warblers in summer. Crossbills are often reported, but difficult to pin down to a particular site. Look for them in sections bearing a good cone crop. In thinned areas Treecreepers feed, Jays breed, and Sparrowhawks dash along the rides with murderous intent. Buzzards soar on broad rounded wings above the woods, and the ominous black shapes of Ravens, croaking and tumbling excitedly in spring, are often seen.

In recently felled zones – there is a vast area in the High Vinnalls region at the moment – Kestrels hunt, and where clearance on this scale has taken place there is always the possibility of Hen Harriers and Short-eared Owls. Nightjars too should find this area suits their requirements. Small birds, especially the two common pipits, Dunnocks, Wrens, Whinchats, Grasshopper Warblers, Whitethroats and Linnets breed here, and Cuckoos are attracted by the abundance of host species.

Deciduous areas offer a much greater variety of bird species. An excellent oak wood can be found alongside the B4361 Richard's Castle road, which extends along the south-west slopes of the Mary Knoll Valley at Overton Common. A small open area still exists between the wood and the beech shaded road. The drumming of the two pied woodpeckers and the yaffling of Green Woodpeckers are features of this site, where Nuthatches and Treecreepers are ever present. Summer visitors include Redstarts, Wood Warblers, Chiffchaffs and other woodland songsters. The same species, along with handsome Pied Flycatchers, are found in the hillside wood at the other end of the valley. There is an attractive open area to the north and west of this wood, falling to the alder-lined stream in the valley bottom, which is often favoured by fallow deer. Birches grow alongside the forest roads in places and are sought out by Siskins and Redpolls in winter. Other birds found in deciduous patches of woodland are Sparrowhawks, Tawny Owls, Woodcocks, Cuckoos, Jays, Spotted Flycatchers, Long-tailed and five other species of tits, all the common woodland songbirds, and finches, including rare appearances by the elusive Hawfinches. Tree Pipits can be seen parachuting in more open areas and along woodland edges.

## Timing

Disturbance is from casual visitors, who do not usually stray far from parked cars, and ramblers who stick to paths and forest roads, causing minimal upset. Forestry operations should obviously be avoided.

Spring and early summer give optimum birdwatching conditions, with residents breeding and migrants arriving. Winter can produce large passerine flocks, especially finches, and large raptors may wander into the area; even a Snowy Owl was seen in the Mary Knoll Valley in the early 1960s.

In spring and summer, early morning trips are worth the effort, and particularly good .for birdsong. Warm days at this time of the year encourage raptors to soar, the mid-day period producing maximum activity. Evenings should not be neglected, with roding Woodcocks, Tawny Owls vociferous, a late hunting sortie from some hungry raptor, and the chance of seeing large mammals in the form of fallow deer, fox and badger. Bats have been encouraged by the provision of bat-boxes in some areas.

## Access

The minor road which leaves the B4361 (A49(T) formerly) on the right, immediately after crossing Ludford Bridge at the south of Ludlow, bisects this area, with Bringewood Chase to the north and the Mary Knoll Valley to the south. There is parking space near the Forestry Commission (FC) office and houses after 1 mile (1.6 km), and there is a nature trail in this area. A large car park has been sited a further 2 miles (3.2 km) west along the same road. There are many forest roads and paths in this region, and the Mary Knoll Valley and High Vinnalls can be reached from here. A more rewarding descent into the valley is by the path from Mary Knoll House, just over ½ mile (0.8 km) east of the car park.

An alternative access point is reached by continuing south along the B4361 for 1 mile (1.6 km), taking the right turn for Richard's Castle to park by the roadside just beyond Overton after a further 1 mile (1.6 km). This gives access to the lower end of the Mary Knoll Valley and Haye Park Wood. Consult the Ordnance Survey map to plan circuit walks. (It is worth noting that in long periods of drought the forest may be closed to visitors to prevent risk of fire.)

## Calendar

*Resident:* Sparrowhawk, Buzzard, Kestrel, Woodcock, Tawny Owl, all three woodpeckers, Nuthatch, Treecreeper, Goldcrest, six tits, resident woodland songbirds, Jay, Raven, resident finches including Crossbill.

*April–June:* Residents breeding, before summer visitors arrive from mid-April, including Cuckoo, Tree Pipit, Whinchat, scrub and leaf warblers, Pied Flycatcher. Spotted Flycatcher and possibly Nightjar from May. Soaring raptors.

*July–September:* Flocks of mixed juvenile tits. Finch flocks forming. Soaring family parties of Buzzards.

*October–March:* Finches including Siskin, Redpoll and occasionally Hawfinch. Wandering large raptors. Crossbill and Raven breeding from February. Chiffchaff arrives late March.

## WIGMORE AREA

### Habitat

This countryside is an area of contrasts where the flat wet meadows south of the River Teme meet the foothills of the Welsh border. The three principal habitats are the river, the wooded hills of Wigmore Rolls and the flat expanse between the two.

The Teme flows generally south-east from Leintwardine before passing below Criftin Ford Bridge, after which it swings in a huge loop to pass through Burrington Bridge and eventually heads north-east through Downton Gorge. The area contains some amazing meanders, especially between the two bridges. Vertical cliffs have formed on the outer bends, and at low water levels pebble beaches are visible where slower currents have deposited them, creating conditions suitable for Kingfishers on the one hand and for waders, wagtails and Dippers on the other. Winter rains bring flooding on either side of Criftin Ford Bridge making the area attractive to wildfowl.

The flat expanse of Wigmore Moor is still labelled Wigmore Lake on some maps, a reminder of the days before the land was drained. The landscape of today is one of small fields, and hedgerows with an abundance of timber, in which willows dominate. The land is low-lying and damp, intersected by drainage ditches, but with very few permanently marshy areas remaining. Some years ago extensive flooding occurred most winters, bringing in good numbers of wildfowl, and the general wet conditions were suitable for waders at other times of the year, but drain clearing has resulted in considerably less flooding in recent years with a corresponding fall in numbers of wetland species.

The extensive woodland to the west of the village of Wigmore is known as Wigmore Rolls and is yet another section of Mortimer Forest. Planted on rolling hill country, the Rolls are a blanket of mainly coniferous trees. Some areas of broad-leaved species can be found, especially in Barnett Wood, the woodland to the south of the lane which leaves Wigmore west.

## Species

With the Teme at lower levels, Grey and Pied Wagtails feed on pebble beaches, from which occasional Dippers base their underwater explorations, with Common Sandpipers in passage periods and possible Green Sandpipers in autumn and winter. Sand Martins and Yellow Wagtails grace the river in spring and summer, but both species have declined in recent years, whilst the brilliant flash of a Kingfisher may be seen negotiating the bends at considerable speed. Grey Herons are regular visitors, and other non-breeding fish-eaters in the area are Cormorants and Goosanders. A favourite gathering ground for several species to feed, bathe, preen and rest is the northern tip of the large meander due east of Criftin Ford Bridge, where the river swells into a pool. Little Grebes sometimes feed here, and flocks of Lapwings and Curlews use the site, whilst wildfowl include Mallard, small numbers of Teal and often large numbers of wintering Wigeon. In times of flooding wildfowl numbers increase and less usual species may occur.

Wigmore Moor, where Teal have bred in the past, and a flock of seven Black-tailed Godwits was noted in the early 1960s, with a Garganey four years later, is a much quieter area these days. Lapwings and Curlews are still present, with a few Snipe in the wetter places and ditches. It is the fields and hedgerows now that provide food, shelter and nest-sites, with large flocks of Fieldfares and Redwings in autumn and winter, and good numbers of finches, including Redpolls, feeding on the seeds of meadowsweet and other weeds growing along the green lanes. Barn Owls, another sadly declining species, may still be seen, and on winter afternoons emerge well before dusk to commence hunting. Other predators are Sparrowhawks, Kestrels and Little Owls, the latter probably nesting in hollow willows. Other species using tree-holes are Stock Doves, Jackdaws, Starlings, Tree Sparrows and occasional Redstarts. Searching the hedgerow timber for insect food are Great Spotted Woodpeckers, Nuthatches and Treecreepers. A limited amount of flooding still occurs in some winters, bringing wildfowl and gulls to the area.

The afforested hills to the west of Wigmore hold good numbers of birds, but the conifer domination leads to a smaller number of species. Goldcrests and Coal Tits prefer these areas, but it is Crossbills that will probably give birdwatchers more pleasure, and the Rolls produce regular sightings. Siskins and Redpolls also feed in the conifers, but seek birch seeds too, and more common finches feed on weed seeds along the edges of forest roads and firebreaks. Great Spotted Woodpeckers and Jays find the conifers to their liking, and in the more open areas Nuthatches and Treecreepers feed on the trunks. There are a few sections of broad-leaves, which give a greater variety of species, with thrushes, scrub warblers, leaf warblers and tits making up much of the breeding population. Sparrowhawks hunt the woods, with soaring Buzzards and Ravens a familiar sight.

## Timing

This is an area of little disturbance, and offers good but unspectacular birdwatching throughout the year. Winter has much to offer, with wildfowl attracted by the floods, flocks of thrushes and finches on the moor, and Crossbills, Siskins and Redpolls in the forest.

The time of day to birdwatch is not so critical as in some areas, but birds are generally more active early in the day, and large soaring birds are best seen around mid-day on warm days.

## Access

The river can be watched from the lane running at a higher level on the northern side, the best point being at the T-junction just to the east of Nacklestone Farm. Approach by car is along the Ludlow to Wigmore road, leaving Ludlow west immediately after crossing Ludford Bridge. After 3½ miles (5.6 km) turn right and continue for a further 2 miles (3.2 km). An alternative approach is from Leintwardine, taking the minor road east from the bridge and following the northern side of the river to reach Nacklestone in 1¾ miles (2.8 km).

To reach Wigmore Moor, take the lane crossing Criftin Ford Bridge south-west to reach Adforton on the A4110 in 1½ miles (2.4 km), turning left here to approach Wigmore. The moor is to the east of this road, and a good access point is the lane opposite the cemetery, just to the north of the village.

For Wigmore Rolls, take the lane leaving the village west at the southern end. This lane, hilly and narrow, reaches the forest in about 1 mile (1.6 km).

## Calendar

*Resident:* Heron, Mallard, Sparrowhawk, Buzzard, Kestrel, Barn Owl, Little Owl, Kingfisher, Great Spotted Woodpecker, Grey Wagtail, Goldcrest, Coal Tit, Nuthatch, Treecreeper, Raven, Tree Sparrow, Crossbill.

*April–June:* Common Sandpiper pass. Summer migrants from mid-April, including Cuckoo, Sand Martin, Yellow Wagtail, Redstart, scrub and leaf warblers. Soaring raptors.

*July–September:* Post breeding flocks of Lapwing and Curlew form. Common and possibly Green Sandpiper pass. Tits and finches flocking.

*October–March:* Cormorant. Winter wildfowl including Wigeon, Teal, Goosander. Snipe, Dipper, Fieldfare, Redwing, Siskin, Redpoll.

# COLSTEY WOOD &
# BURY DITCHES

Map 37
OS map 137

## Habitat

The south-west corner of Shropshire contains many modern coniferous plantations, which clothe the slopes and even the summits of some of the numerous hills. Most are accessible by track and path, but usually to a very limited extent. One of the better ones to visit is the large area of woodland referred to as Colstey Wood, and containing the ancient hill fort of Bury Ditches, which is situated to the east of the A488 between Bishop's Castle and Clun. Formerly completely planted with conifers, the site of the fort is now cleared, with the rings of the ramparts clearly defined. In spring it abounds with pipits and it is an excellent vantage point from which to see soaring raptors, in addition to providing superb views of the Welsh Marches.

The wood can be explored from several waymarked trails of varying lengths, set out by the Forestry Commission (FC), so that the bird-watcher has options from a very short walk to a more adventurous ramble of 4 miles (6.4 km) or more. Although at first glance the wood may appear to be almost completely coniferous, there is actually a surprising amount of broad-leaved timber, and the longer walk takes in a good variety of woodland habitats including a small pool and a woodland stream.

## Species

Although the conifers contain a selection of birds including Wood-pigeons, Jays and Chaffinches, the specialists here are Goldcrests, Coal Tits, Siskins in winter, and Crossbills. The latter are recorded here more often than at any other site in Shropshire.

In the deciduous and mixed areas of the woods, a greater variety of birdlife is seen. Sparrowhawks breed, often siting their nest in a larch isolated amongst broad-leaved trees, and their prey includes most of the common woodland passerines. All the typical species can be found, such as Wrens, Dunnocks, Robins, Redstarts, scrub and leaf warblers, Spotted and Pied Flycatchers, six species of tits, Nuthatches, Treecreepers, Greenfinches and Bullfinches. All three woodpeckers occur here, along with Turtle Doves, Jays, Jackdaws, and the crepuscular Woodcocks and Tawny Owls.

Along the woodland edges, in clearings, rides and the more open areas, Tree Pipits are numerous in summer, with Siskins and Redpolls being attracted by birch seeds during winter months. Where thistles and other desirable weeds grow, dancing flocks of Linnets and charms of brilliant Goldfinches are numerous in late summer and autumn.

Bury Ditches provides a glorious panorama, and is also a commanding observation point from which to see large birds in flight. Soaring Sparrowhawks, Buzzards, Kestrels and Ravens are commonplace, and other large raptors may be seen, including rare non-breeding season visits by Red Kites. Both common pipits breed here, and Crossbills regularly overfly the fort when moving between feeding areas. Cuckoos are attracted by the pipits, whilst Swifts, Swallows and House Martins find a good food supply. Wheatears, upright in stance, and further betrayed by flashing white rumps, occur on passage.

*Sparrowhawks can often be seen in the Clun Forest*

## Timing

The area is mainly visited by ramblers and casual walkers, with summer weekends, especially Sundays, bringing greatest numbers. However, this part of the county does not attract the hordes of tourists as do the more publicised areas. The only other likely cause of disturbance is from forestry operations.

Spring and summer are the best times to visit, although there is usually enough interest at any time of year to warrant a trip.

Early morning is the time for greatest bird activity, but raptors are best seen around mid-day in warm weather, after the formation of thermals, which encourage soaring. Evening trips are also worthwhile, with Woodcocks, Tawny Owls and late hunting predators the attraction.

## Access

The FC has provided a car park at the western end of Colstey Wood. Access is from the A488, 2¼ miles (3.6 km) north of Clun or 3½ miles (5.6 km) south of Bishop's Castle. There is a display board here with a map showing the various colour-marked trails.

An alternative starting point, and much closer to the Bury Ditches hill fort, is from the unclassified narrow lane running north from Clunton village to join the B4385 at Brockton. Motorists using this lane in either direction should take the eastern of the two roads through Brockton, the ford on the other being unsuitable for family cars. There is ample parking space just off the lane ½ mile (0.8 km) south of Lower Down.

## Calendar

*Resident:* Sparrowhawk, Buzzard, Kestrel, Woodcock, Tawny Owl, three woodpeckers, Nuthatch, Treecreeper, resident thrushes, six species of tits, Goldcrest, Jay and common crows, Raven, resident finches, including Crossbill, Yellowhammer.

*April–June:* Summer migrants from mid-April, including Cuckoo, Tree Pipit, Redstart, scrub and leaf warblers, Pied Flycatcher, Swallow, House Martin. Turtle Dove, Swift and Spotted Flycatcher from early May. Roding Woodcock. Soaring raptors.

*July–September:* Residents and summer migrants still breeding. Family parties of raptors. Flocks of mixed juvenile tits. Finch flocks, especially Goldfinch and Linnet.

*October–March:* Raptors, Siskin, Redpoll. Raven courtship from January. Resident species, for example, Meadow Pipit, moving back to territories at end of period. Chiffchaff late March.

# CLEE HILLS & CATHERTON COMMON

## Habitat

The Clee Hills are made up of the twin summits of Brown Clee and Titterstone Clee, the eastern slopes of which merge into the extensive unenclosed Clee Hill Common. The hills are aligned north to south with the summit of Titterstone Clee less than 6 miles (9.6 km) from the Hereford and Worcester border. Consisting of Old Red Sandstone, capped by harder dolerite which has protected underlying coal through the ages, the hills have suffered at the hand of man and show the scars.

The northern hill is Brown Clee, its major summit of Abdon Burf the highest point in Shropshire at 1,772 ft (540 m). Almost 1½ miles (2.4 km) further south the minor summit of Clee Burf is 100 ft (30 m) lower. Both summits carry masts, the price we pay for modern technology, but at least they make good perches for Kestrels. Spoil heaps and derelict buildings near Abdon Burf are gradually weathering into the landscape.

Brown Clee is a huge sprawling hill containing a large variety of habitats. The western flanks are open, covered by acid grassland, some extensive patches of gorse and bracken. The plateau contains heather moorland south of Abdon Burf, and wet heath south of Clee Burf. Several pools dot the high ground. The woodlands lie along the eastern slopes, a large proportion consisting of modern coniferous plantations, but with generous areas of oak, birch scrub and hawthorn scrub. There is also some fine open parkland. The nature trail in the north-east is set out in mainly coniferous woods. The streams are rather insignificant, being mainly hidden in woodland; even the one flowing west from the col to Cockshutford is hidden from sight by hedges and trees. Nordybank, a well-defined hill fort, sits on the end of a spur of high ground running west from Clee Burf.

Approached from the west, Titterstone Clee is impressive. The northern and western slopes, a mixture of heath-grasses, bracken, some scrub gorse and rushy wet flushes, rise to merge into boulder scree leading to the rocky eminence of the Giant's Chair near the summit of 1,749 ft (533 m). Approach from the south is disappointing, the scars, spoilheaps and derelict buildings of former quarrying operations giving an air of desolation, and the science fiction installations of the Civil Aviation Authority radar station, looking like giant tee-ed up golf balls, are very prominent. Do not be put off by this despoliation of the hill, there is much left to admire. Once clear of the eyesore, the eastern slopes swell into a wide flat plateau of wet heath, with a few dragonfly-haunted small pools, patches of soft rush and elevated dry islands of gorse. Finally the hill runs down into Catherton Common. Streams are not a distinctive feature of the landscape, but the one flowing north to pass beneath the road near Cleeton St Mary attracts occasional Grey Wagtails. The working quarry near Clee Hill village is far enough removed from the hill as to cause no disturbance to birdwatching in the more favourable areas.

Managed by the Clee Hill Commoners Association, Catherton Common forms a substantial part of the unenclosed common land to the east of Titterstone Clee. It is 300 acres (120 ha) of heathland, on which heath-grasses, heather and bilberry dominate, with patches of gorse and birch scrub. Water filled depressions, fringed with willow and aquatic vegetation, are the present day evidence of mediaeval coal mining by bell-pits. The wet areas attract dragonflies, insectivorous sundew and butterwort, with a breeding population of Reed Buntings. Adders are recorded on the common, making stout footwear an essential, especially on hot sunny days when they bask in the open. The smallholdings scattered about the common have tall sheltering hedges which add diversity to the habitat as do the coniferous plantations along the south-east boundary.

## Species

Buzzards and Kestrels are found throughout the area, and families of both species may be found hovering in line abreast, using the updraughts along western facing slopes in late summer. Other birds common to all areas are Snipe in wet heath, Cuckoos with their commonest host species, Meadow Pipits, and Tree Pipits along woodland fringes and scrub areas. Large flocks of Swifts gather to feed on the summer insect bonanza, with good numbers of Swallows and House Martins, whilst fruit in the form of bilberries attracts Mistle Thrushes. All areas contain some gorse scrub and bracken, suitable for Whinchats, Stonechats, Linnets and Yellowhammers. Jays, often seen in floppy flight heading for the woods, Magpies, and the common black crows are all numerous. Reed Buntings breed in the wet heaths, and many common passerines are found in the fringe areas.

*The Raven's croaking call and thrilling flight are symbolic of wild hills such as Clee*

Ravens, the scarce Merlin, and breeding Wheatears are found on the higher ground. Ring Ouzels may occasionally breed, but are more frequently seen in numbers on spring and autumn passage. Curlews are surprisingly scarce on the hills, but breed in surrounding fields.

Brown Clee, with a larger variety of habitats, provides the longest species list. Woodcocks and Tawny Owls breed in the broad-leaved woods, which are also home to Sparrowhawks, the three woodpeckers, and most woodland passerines, including Wood Warblers, Pied Flycatchers, Nuthatches and Treecreepers. Birch scrub also holds the woodpeckers, with Redstarts, tits, warblers, Siskins and Redpolls. Crossbills are now seen regularly in the coniferous plantations, with other small passerines including the numerous Goldcrests and Coal Tits. The exciting Goshawk is occasionally sighted in these conifers. The well-wooded parkland usually shows evidence of extensive Pheasant rearing on the estate, and Green Woodpeckers are regulars. Sadly, the Woodlarks once found here have never re-colonised after being wiped out by a dreadful winter in the early 1960s. The high altitude pools on Brown Clee are shunned by swamp warblers, but Teal, Mallard and Tufted Duck are found on them. Passage Redshanks and Green Sandpipers have been known to call at the more open pools near Abdon Burf. Boyne Water, at 1,470 ft (448 m), holds the highest nesting Coots in Shropshire. A few Red Grouse are found in the heather south of Abdon Burf.

The boulder scree and rocks of Titterstone Clee are visited by Peregrines, and often used as look-out posts by Kestrels and Ravens. Snow Buntings have been seen here in late winter, and Dotterels recorded on passage. Pied Wagtails haunt the old quarry buildings, but more exciting are the occasional Black Redstarts that show up in spring.

Catherton Common has been neglected by Nightjars for some years now, but the habitat is suitable and re-colonisation is a much hoped for possibility. Curlews and Skylarks breed on the heath, with Short-eared Owls hunting over the heather in some winters. Mallard and Reed Buntings nest in damp areas and around the pools, with birch scrub providing food and shelter for Great Spotted Woodpeckers, Willow Warblers, tits, Siskins and Redpolls. The woodland on the south-east boundary of the common holds Sparrowhawks, Woodcocks and Tawny Owls. Little Owls can be seen in the daytime, perched on posts and wires, and the sheltering hedges of the smallholdings provide nest-sites, song posts, food and shelter for the common passerines.

## Timing

Although attracting many visitors, the area is not as disturbed by the summer hordes as some of the more accessible Shropshire hills. Serious walking is required to get the best from this hill country, and most tourists are content to walk around the nature trail on Brown Clee, or drive to the old quarries to enjoy the view from Titterstone Clee. In conditions of good visibility the Clee Hills offer panoramic views which are unsurpassable, and as the same conditions usually produce good birdwatching, opportunities to visit at these favourable times should not be missed. Winter weather makes the higher regions inhospitable, although Ravens, inevitable crows and occasional raptors will be present. Late spring will give the maximum number of species.

The best time of day depends on observers' requirements. Early morning will give greatest bird activity, but soaring raptors require thermals and are best seen around noon. Evenings can also be excellent, with raptors looking for a late kill and nocturnal species beginning to stir. Take the necessary precautions if you expect to be on the hills when darkness falls. Catherton Common can be quite magical at dusk, with roding Woodcocks, drumming Snipe, crowing Pheasants, hooting Tawny Owls, yelping Little Owls, the rush of Mallard wings through the gathering gloom, and the anxiety calls of Curlews, disturbed perhaps by some prowling fox.

## Access

Brown Clee is surrounded by a maze of narrow winding country lanes which form a network between the B4364 Bridgnorth to Ludlow road, and the B4368 Morville to Craven Arms road. The former passes below the eastern slopes of the hill, but the only access point from the road is along the wide track leaving due north 1¼ miles (2 km) south of Burwarton, where there is space to park. To approach from the west, leave the B4364 west for Stoke St Milborough to park at The Yeld, ½ mile (0.8 km) south of Clee St Margaret, or continue through the village to park at the north-west of Nordybank Hill Fort. Roadside parking 1¼ miles (2 km) west of Cleobury North at Cleobury North Liberty gives access from the north.

For Titterstone Clee leave the A4117 Ludlow to Cleobury Mortimer road along a minor lane (signposted Dhustone) ¾ mile (1.2 km) west of Cleehill village. Fork left on the hill to park on a wide flat area beside the old quarry. To approach from the east, take the minor road from Foxwood, 2¼ miles (3.6 km) east of Cleehill and park at the bottom of the track from Magpie Hill, or at Cleeton St Mary.

Catherton Common also lies north of the A4117 and can be approached from Foxwood. Minor roads from Hopton Wafers and Cleobury Mortimer, both on the A4117, also lead to the common.

## Calendar

*Resident:* Sparrowhawk, Buzzard, Kestrel, Red Grouse, Woodcock, Little Owl, Tawny Owl, the three woodpeckers, Goldcrest, Nuthatch, Tree-creeper, Jay and other common crows, Raven, Reed Bunting and other resident passerines.

*April–June:* Soaring Buzzards, passage Dotterel (Titterstone Clee), drumming Snipe, roding Woodcock, breeding Wheatear, Ring Ouzel (passage), occasional Black Redstart in old quarry buildings. Cuckoo and other summer migrants from mid-April. Crossbill families.

*July–September:* Raptor families hunt together, soaring parties of Buzzards, large feeding flocks of Swifts (until mid-August), Mistle Thrush flocks, passage Dotterel (scarce).

*October–March:* Peregrine (scarce), Curlews return to breeding grounds in March, vagrant Short-eared Owl, Fieldfare, Redwing, Raven courtship, finch flocks including Siskin and Redpoll, Crossbills breeding from February, occasional Snow Bunting (Titterstone Clee), resident passerines move to high ground in March.

# CHELMARSH RESERVOIR & THE RIVER SEVERN

Map 40
OS map 138

## Habitat

The comparatively new reservoir at Chelmarsh — it was constructed in the early 1960s — lies 3½ miles (5.6 km) south of Bridgnorth. The ¾ mile (1.2 km) stretch of water is aligned from north-west to south-east, and at the southern, deep end, is about ½ mile (0.8 km) wide. The northern end, much narrower and comparatively shallow, is separated by a causeway carrying a public bridleway from a tract of marsh, 250 yds (230 m) long and 100 yds (93 m) wide.

This marsh is bisected along the longest axis by a drainage ditch, and an attempt to reclaim the area some years ago by excavating further ditches appears to have failed. There is a well-established growth of common reed spreading on one side of the drain, and a patch of reedmace and some willows on the other, but the dominant plant is great willow-herb, and in summer the marsh is ablaze with the deep rose flowers. The water table is very high, resulting in small areas of surface water, especially in wet periods, and it is a strong possibility that the marsh will be managed to improve the ornithological value in the near future.

The reservoir is surrounded by a variety of habitats. Alders have been planted along much of the water's edge, rough grassland lines the north-east shore, and the opposite shore supports dense willow growth to the south, which gives way to thorn scrub further north. Chelmarsh Reservoir is owned by the South Staffordshire Water Authority (SSWA). A bird-rich mixed deciduous wood lies outside the SSWA property, but adjacent to the boundary fence in the south-west. The dam embankment is grassed, but kept short by mowing or occasional grazing. The area to the south-east of the embankment is a mixture of rough grasses, a small outlet stream, and mixed woodland which includes some yews, favourites with roosting Tawny Owls.

There are no feeder streams into the reservoir, the water being pumped from the River Severn which is a mere ½ mile (0.8 km) distant. The great advantage of this is to keep at least a section of the water unfrozen in severe weather when most other pools in the Midlands are icebound.

The close proximity of the reservoir to the Severn, with the woods and pools of Dudmaston Hall to the east of the river, creates a varied area within which birds can move to suit weather conditions and food supply. In freezing spells waterfowl can leave the pools to join the river, but when the latter runs discoloured and bank-full they can reverse the move.

This reach of river is rather straight, with the water flowing deep and fast on either side of Hampton Loade. There is no spectacular flooding, but in prolonged wet spells some riverside fields become waterlogged and hold shallow pools which attract a few wildfowl, Lapwings and gulls. Pasture and arable fields border the river, but the extensive mixed woodland of Long Covert closes on the east bank, and the area is dotted with smaller coverts. The confluence of the Mor Brook with the parent river creates an interesting habitat, the water widening here and showing a few pebbly beaches at low water levels.

# Species

From the newly flooded sterile stretch of water, Chelmarsh Reservoir has gradually built up into a first-class wetland habitat giving exciting birdwatching. It is an excellent winter wildfowl site, with both winter swans appearing, although scarce, and the hundreds of Canada Geese being joined by small numbers of wild Pink-footed and White-fronted Geese, feral Greylag and Barnacle Geese, and Snow Geese escapees. Ducks present in good numbers are the ubiquitous Mallard, Wigeon, Teal, Shoveler, Pochard and Tufted Duck, with smaller numbers of Goldeneye, Goosander and Ruddy Duck. There are annual sightings of Shelduck, Gadwall, Pintail and Scaup, with Long-tailed Duck and Common Scoter most years, and scarcer ducks on rare occasions.

Winter also brings other waterfowl to join the resident but non-breeding Great Crested Grebes, with a few Little Grebes each year and scarce Red-necked, Black-necked and Slavonian Grebes. Cormorants in good numbers, and an occasional Shag, share the water with uncommon divers, the Red-throated Diver being the most frequent visitor. Herons fish the reservoir throughout the year, having formerly nested in the adjoining wood, and they still use the old nests as platforms to stand on during resting periods.

Being supplied from the river, the water level does not fall during dry spells, hence wader passage has not been a strong feature of the reservoir, due to lack of suitable shoreline. A few species occur annually, feeding along the water's edge on the concrete dam, or staying briefly where minute areas of sand show in one or two places. These include Oystercatcher, Dunlin, Redshank, Greenshank and Green Sandpiper, with Common Sandpipers quite numerous in some years. Terns are also regular passage visitors, usually Arctic and Common Terns, but also the delicate Black Tern, with rare visits from Sandwich Terns.

An exciting passage migrant is the Osprey, which usually appears in the post-breeding period as the birds move leisurely south. Their visits are becoming more frequent and they should be looked for on dead trees in surrounding hedgerows, as they tend to sit immobile for long periods, despite the attention of Magpies and other crows.

The open water of the reservoir is an ideal gull roost, with the most numerous species being Black-headed and Lesser Black-backed Gulls, which usually give a combined figure of up to 12,000 birds in December and January. Herring Gulls only reach a few hundred, with small numbers of Common and Great Black-backed Gulls. Annual sightings of rarer gulls now occur, with first arrivals being Little Gulls in August, followed by Mediterranean Gulls from October. Iceland and Glaucous Gulls usually arrive after the turn of the year, and in two recent winters Ring-billed Gulls have been reported in the same period. The graceful Kittiwake shows up in most winters, sometimes as late as April.

Two other very different winter visitors are the Kingfisher and Dipper, which leave their more usual river haunts to seek food in the reservoir and outlet stream respectively. The stream also attracts Grey Wagtails.

The marsh holds breeding Grasshopper, Sedge and Reed Warblers, Whitethroats and Reed Buntings. A pair of Snipe breed regularly, but are mainly wintering birds in some numbers, along with one or two Jack Snipe. Water Rails also occur at this time, often feeding in the drain, and Stonechats are seen, almost always in pairs. An exciting newcomer recently was a Cetti's Warbler. The marsh is a regular hunting ground for

Kestrels, and Barn Owls were formerly familiar visitors. The recent excavation of a shallow scrape in the marsh has created a new habitat suitable for waders, and future years should see increased bird activity in the area.

The woodland adjoining the reservoir holds all the expected birds, including Sparrowhawks, Woodcocks, Turtle Doves, Tawny Owls, the three woodpeckers, Nuthatch, Treecreeper and a wide selection of woodland songsters. Even a Firecrest visited the wood on one occasion. Many of these birds are also found in the scrub, especially warblers, with Fieldfares and Redwings arriving for the berry harvest in autumn.

Many species are seen generally around the area, or overflying, and these include both partridges, Lapwings and Curlews. Little Owls are resident, and Short-eared Owls, which hunt the grassland, are fairly frequent visitors. Several Cuckoos are usually vociferous on arrival, whilst Sand Martins and Swallows make this a first port of call before dispersing to breeding sites.

Many species found at the reservoir also occur on the river. Herons are ever present, as are Kingfishers except after harsh winters. Birds breeding along the river are Mallard, Moorhens, Sand Martins, Sedge Warblers, Reed Buntings and Yellow Wagtails in the bordering fields.

Little and Great Crested Grebes move onto the river after breeding, with Grey Wagtails and Dippers moving down from the tributary streams. Mute Swan families also appear at this time, and probably breed on eyots. Birds coming in as winter visitors include Cormorants, Pochard, Tufted Duck, Goldeneye and Goosander.

The Severn also carries passage migrants, including the terns seen at the reservoir, with Dunlins, Redshanks, Greenshanks, Green and Common Sandpipers the likeliest waders. Snipe feed along the river in winter.

If riverside fields are holding flood pools, they attract swans including scarce Bewick's and Whooper Swans, and surface feeding ducks, mainly Wigeon, Teal and Mallard, but also Pintail and Shoveler. Geese feed on the arable fields between the reservoir and river, the species being as listed above.

The woods hold good numbers of birds and there is usually some activity to arouse interest, such as the mobbing of a Cuckoo or a family party of Jays crossing from wood to wood. Sparrowhawks are regularly seen soaring or hunting, but Buzzards are strangely scarce, although they occupy similar habitat both north and south of this reach. The area is very good for wintering flocks of passerines, with tits, Fieldfares, Redwings, Tree Sparrows, Chaffinches, Bramblings, Greenfinches, Goldfinches, Yellowhammers and Reed Buntings often very numerous.

## Timing

Winter is the best season at the reservoir, especially from mid-November on. The water tends to be devoid of birdlife in high summer. April to mid-May and August to mid-September should produce passage species.

There is disturbance at the reservoir from sailing and angling, with the latter sport a great favourite on the river. An early visit will avoid the sailors but not the anglers. Weekends are the worst times. Visits following strong westerly gales often pay dividends in winter and, contrary to most lakes, the reservoir should be visited during freezing spells.

The position of the sun can effect birdwatching across large sheets of water, and this should be checked on arrival before deciding on a route. There is a small hide at the corner of the bay in the south-west of the reservoir, which faces north, making it a good spot at any time of day.

## Access

To reach the area from Bridgnorth, take the unclassified road from the sharp bend near the railway station and travel south. After 2¾ miles (4.4 km) roadside parking just before reaching a sharp bend below a railway bridge will give access to the river along the Mor Brook. To reach Hampton Loade, continue along this road, turning left at Sutton into a no-through-road which leads to the river. The Severn can be walked for miles in both directions from these spots, and the Hampton Loade starting point has the novelty of a foot ferry should one care to walk the opposite bank.

Access to Chelmarsh Reservoir is by permit only, issued to members of the Shropshire Ornithological Society (SOS). The access road is from Hampton on the no-through-road. Non-permit holders can view the reservoir and marsh from the public bridleway which starts from near Chelmarsh Church, or from various public footpaths on the north, east and south. A public observation area is provided beside the boundary fence at the southern end.

Birdwatchers who enjoy a good ramble can choose a circuit around the complete area.

## Calendar

*Resident:* Great Crested Grebe, Grey Heron, Mute Swan, Canada Goose, Mallard, Tufted Duck, Sparrowhawk, Kestrel, Snipe, Woodcock, Little Owl, Tawny Owl, Kingfisher, three woodpeckers, Grey Wagtail, Dipper, Goldcrest, Nuthatch, Treecreeper, Reed Bunting.

*April–June:* Summer visitors from mid-April, including Cuckoo, Sand Martin, Yellow Wagtail, Grasshopper, Sedge and Reed Warblers, scrub and leaf warblers. Turtle Dove, Swift and Spotted Flycatcher from early May. Passage waders including Oystercatcher, Dunlin, Common Sandpiper, and passage terns, including Black Tern.

*July–September:* Possible Osprey. Passage waders, including Dunlin, Redshank, Greenshank, Green and Common Sandpiper, and passage terns, including Black Tern. Occasionally Little Gull.

*October–March:* Occasionally divers, especially Red-throated Diver, grebes, including rare Black-necked, Red-necked and Slavonian Grebes. Cormorant, small numbers Bewick's and Whooper Swans, Pink-footed and White-fronted Goose, ducks including Wigeon, Gadwall, Teal, Pintail, Shoveler, Pochard, Scaup, Long-tailed Duck, Common Scoter, Goldeneye, Goosander. Water Rail, Jack Snipe. Gull roost, including scarce gulls. Short-eared Owl, Stonechat, winter thrushes, finch flocks, including Brambling.

# MUCH WENLOCK AND
# WENLOCK EDGE

**Map 41**
OS maps 127 & 138

## Habitat

The internationally famous Wenlock Edge, a low, wooded limestone escarpment, runs some 20 miles (32 km) south-west from Ironbridge to Craven Arms. Originally clothed in ash and oak, with a mixture of other trees, it has been converted to conifers in some areas during recent decades. A beneficial development has been the acquisition of substantial areas by the National Trust (NT), with a management plan of reconversion to ash woodland. The woods lie on the steep scarp slope, rarely exceeding ¼ mile (400 m) in width, and the gentle dip slope is mainly given over to cereal production, with some pasture for sheep and cattle. South of Much Wenlock, as far as Hilltop, the Edge is much scarred by quarrying, but this does create some interesting waste areas. The occasional rock outcrops, usually well hidden by trees, are not extensive enough to attract cliff-nesting birds, although Kestrels, Stock Doves and Jackdaws do take advantage of disused quarry faces in which to breed. There is little running water, and no standing water apart from the occasional small pond, but one substantial stream, known in different localities as Hughley Brook or Sheinton Brook, flows below the Edge, passing through Harley, the amusingly named Wigwig and Sheinton before joining the Severn. The disused railway track between Much Wenlock and Farley is now an attractive habitat for hedgerow birds, offering both food and shelter to newly arriving migrants in both spring and autumn. The whole area is worth exploring for woodland species, but is also rich in interesting limestone flora and good for butterflies. The early riser may be rewarded with sightings of foxes and fallow deer, whilst the late evening watcher may also see badgers.

## Species

The Ordnance Survey map shows an area of woodland bisected by Sheinton Brook lying 2 miles (3 km) north of Much Wenlock and labelled The Springs. This is known locally as Bannister's Coppice. It is about 220 acres (90 ha) of deciduous woodland, mainly oak with birch and hazel, and is accessible by public right of way. Sparrowhawks and Buzzards are regulars here, and there has been one winter sighting of a Red Kite. This woodland provides roding Woodcock in the breeding season, but they may also be disturbed on winter days if they roost near the paths, and all three woodpeckers, Nuthatches and Treecreepers. Many small passerines breed here, and in spring the wood is loud with birdsong, Garden Warbler, Blackcap and Wood Warbler being very prominent. Tawny Owls are abundant in the area and the coppice echoes with Cuckoo calls in May. Mallards and Dippers breed along the brook and Grey Wagtails can be seen feeding on shallow boulder strewn stretches and pebbly margins. The presence of fallow deer is betrayed by many slots and it is not unusual to see them.

The excellent track leading from Much Wenlock to Blakeway Coppice was once the coach road to Shrewsbury. This area is the Wenlock Edge and in spring and summer the track is bordered with wild flowers,

*Woodcock may be disturbed on winter days, but are more likely to be seen roding in the breeding season*

butterflies abound, and the hedgerows provide a suitable habitat for Lesser Whitethroats, Whitethroats, Linnets and Yellowhammers, although the first named are very scarce. Old quarry workings on the right, now overgrown with scrub, mainly hawthorn, produce an abundance of berries to the delight of Redwings and Fieldfares, whilst the seeds of ground vegetation attract Greenfinches, Goldfinches and Bullfinches. Turtle Doves croon here through summer afternoons.

The woodlands on this part of the Edge are owned and managed by the NT who have cleared footpaths, provided marker posts and generally made the area much more accessible. Coniferous sections contain Goldcrests and Coal Tits, with a good mixture of species in the broadleaved woods. Larger birds found here are Buzzards, Sparrowhawks, Woodcocks, Woodpigeons, Cuckoos, Jays and Carrion Crows, whilst Chaffinches and Willow Warblers are the most numerous small passerines in spring and summer, and all the English tit species are present through the autumn and winter, sometimes in large mixed flocks. Great Spotted Woodpeckers are the commonest of their family, drumming noisily in late winter and early spring, and Green Woodpeckers are often seen ground-feeding along the woodland edge. Lesser Spotted Woodpeckers are uncommon, but may be overlooked. Both Nuthatches and Treecreepers are to be found here in all months. Tawny Owls are more often heard than seen in the woods and Little Owls call noisily in spring from hedgerow timber. Both are most frequently seen by road users, perched on wayside posts and wires at dusk.

Do not be put off by the extensive quarry workings bordering the upper boundary of the woodlands. Stock Doves breed here, large flocks of Jackdaws perform thrilling aerobatics and Kestrels can be seen at close quarters as they glide below the observer. Often, on summer evenings, large numbers of Swifts, Swallows and House Martins feed above the quarries, and these have attracted a Hobby on a few occasions.

Large birds appear to use the Edge as a flight route, perhaps even as a navigational aid. No doubt on some days the updraught from the steep escarpment could provide an economical saving of energy, and all the common raptors, swans, geese, Herons, gulls and Ravens have been noted following the line of the Edge.

The southern stretches of Wenlock Edge should not be neglected. The woods here are less visited by tourists than those in the north, but adequate rights of way exist to allow full exploration. For a first visit the Edge Wood Nature Trail near Westhope, 11 miles (17 km) south-west of Much Wenlock, cannot be bettered. There is parking space here and a picnic site. Longer walks into Wolverton Wood and Harton Wood can be taken from this same starting point. All the species mentioned above may be encountered in these woodlands, along with Pied Flycatchers in spring and summer, with large flocks of Redpolls in late winter. Good numbers of Wood Warblers are usual here, especially where oaks dominate.

## Timing

For small passerines, especially woodland species, March to May for residents and April to May for summer migrants are the best times of the year, before the trees come into full leaf. Early morning is the best time of the day. The development of tourism in recent years has brought more disturbance to late morning and afternoon birdwatching. For Woodcocks and owls, dusk is the only time to be sure of success. Soaring raptors, especially Buzzards, are best looked for around noon on warm days.

## Access

To reach Bannister's Coppice, park in Homer, 1½ miles (2.4 km) by road from Much Wenlock. Take the first turn left after leaving the town northwards on the B4378. The footpath starts between two bungalows, and appears to pass through the garden of one of them, the stile being sited alongside the drive gate.

Wenlock Edge can be reached from the NT car park at the south-west corner of the town, on the northern side of the B4371 just before leaving the built-up section of the road. Continuing along this road there are other good access points between the junction of the road to Hughley and the Wenlock Edge Inn at Hilltop.

To reach the southern woods of Wenlock Edge take the B4378 south-west from Much Wenlock, joining the B4368 at Shipton. Continue ½ mile (0.8 km) beyond the B4365 turn to Ludlow and take the first road on the right into Siefton Batch, through Westhope, to the picnic site and Edge Wood Nature Trail, parking on the right soon after reaching the woodland.

## Calendar

*Resident:* Sparrowhawk, Buzzard, Kestrel, Tawny and Little Owls, the three woodpeckers, Pied Wagtail, Dipper, all common woodland passerines, all common crows, common finches and Yellowhammer.

*April–June:* Roding Woodcock. Summer migrants from mid-April, including Cuckoo, hirundines, scrub and leaf warblers. Redstart and

Pied Flycatcher mainly passage. Swift, Turtle Dove and Spotted Flycatcher from early May. Soaring raptors.

*July–September:* Occasional Hobby. Mixed flocks of juvenile tits. Finch flocks reforming. Jays carry acorns.

*October–March:* Curlews to breeding fields in early March. Large flocks of Fieldfares and Redwings except December and January. Siskin and Redpoll from January. Chiffchaff from end of March.

The underlying rocks in the area occupied by the new town of Telford are mostly middle and upper coal measures of the Carboniferous period. Apart from coal, sandstone and limestone are present, outcropping in the south, with various clays such as fireclay. The coming of the Industrial Revolution and with it the practice of intensive mineral extraction, resulted in the creation of many flat-topped mounds, caused by the dumping of waste material; and the mining of open cast coal and clay, together with balancing reservoirs needed for the canals, left the area dotted with numerous pools. During the century of decay which followed the industrial boom, nature, as always, clothed the eyesores with a great variety of vegetation, which in turn provided ideal conditions for wildlife. It was on this semi-derelict area that Telford was built.

From the start the aim was to preserve a balance between built-up areas and open spaces, retaining some of the habitats for wildlife conservation, and creating others to serve as recreational areas in addition to their wildlife interest. The result is a mosaic of semi-wild habitats, sandwiched between residential and industrial estates, shopping centres, and a network of roads, which contain over 140 bird species, about 80 of which breed, with many species of mammals, invertebrates and flora. There are over 40 spaces with natural history interest, and nearby areas of great ornithological value in The Wrekin, The Ercall, Benthall Edge and the Severn Valley.

## TELFORD

### Habitat

It would be impracticable to describe or map the entire area in detail, and the best plan is to visit the Stirchley Grange Environmental Interpretive Centre where many leaflets describing all aspects of the new town are available. The Ranger Service is based at the Centre, and there is usually something of interest in the offing, such as guided walks, talks, or visits to various places of interest in the town.

Probably by far the most valuable habitat preserved is the ancient woodland and formerly coppiced woods of the Severn Gorge. Secondary woodland, mainly birch, ash and oak, can be found at Lightmoor, Dawley, Donnington Wood and in the Town Park, these same areas also containing acidic grassland and heath, whilst modern plantations will be found at Jiggers Bank and Nedge Hill. A few unimproved meadows — a rarity these days — still exist at Lightmoor and in the Severn Gorge. More than 20 pools are located within the new town boundary, and although some are in built-up areas, others are in semi-wild habitat with a rich variety of wildlife. Some marshy terrain still survives at Donnington Wood.

Some sites are extensive and hold a mosaic of habitats, such as Lightmoor with its birch and oak woods, grassy patches, heather and small pool. The Town Park, immediately south of the Town Centre, is

1¾ miles (2.8 km) long and contains no less than nine pools, acres of woodland, and mounds covered with grass and heather.

## Species

The woodlands hold all the expected species. Sparrowhawks are widespread, and there is an abundance of year round small passerine prey available in the form of resident thrushes, tits including Long-tailed Tits, Wrens, Dunnocks, Nuthatches and Treecreepers. In summer the numbers are augmented by Wood Warblers, Whitethroats, Garden Warblers, Blackcaps, Redstarts and occasional Pied Flycatchers, with Tree Pipits in fringe areas. Larger birds are Green and Great Spotted Woodpeckers, Jays, Stock Doves and Tawny Owls. Coal Tits, Goldcrests and very occasional Crossbills are found in the plantations. Linnets and Goldfinches are regular in scrubby areas, and birches attract Siskins and Redpolls in winter. Nightingales, formerly heard annually in the Severn Gorge, have been absent in recent summers, but still remain a possibility.

Birds of wetland habitats are well represented. Herons, Kingfishers, Grey Wagtails and Dippers are all regular on the Severn in the Ironbridge Gorge, with many other species passing along the river on migration. The numerous pools hold good wildfowl numbers in winter, with Priorslee Flash, Priorslee Lake, Holmer Lake and the Town Park pools taking the greatest numbers. Mallard, Tufted Duck and Pochard are the commonest species, with Wigeon, Teal and other ducks in lesser numbers. Occasionally, scarcer birds are seen: a Long-tailed Duck in November 1985 and Common Scoter on a few occasions being examples. Other waterfowl on the pools are Little and Great Crested Grebes, Herons, Mute Swans, Canada Geese, Moorhens, Coots, Kingfishers, and breeding Sedge Warblers and Reed Buntings.

## Timing

The woodlands are best visited in spring and early summer, although wintering Siskins and Redpolls feed actively in birches and alders, especially after the turn of the year. The pools have good numbers of breeding birds, but really come into their own with the arrival of northern wildfowl from mid-October.

Angling is permitted on most pools, and sailing on Holmer and Priorslee Lakes, with many open spaces used for other recreations. Nedge Hill is a favourite picnic spot, being an excellent viewpoint in addition to the woodland making ideal family walking.

Most of these activities can be avoided by an early morning birdwatching session, although on most pools many of the birds have become quite tolerant of nearby human presence.

## Access

The rapid development of Telford, and the continuing construction of new routes within the area means that maps available in shops rarely show an up-to-date road plan. The large scale street plan produced by Telford Development Corporation is indispensible.

Stirchley Grange Environmental Interpretative Centre is to the west of Queensway (A442) from which it can be reached by leaving at the Stirchley Interchange, and proceeding via Stirchley Avenue, Randlay Avenue and Grange Way. The Centre is at the eastern end of Grange Pool and has a car park. Queensway can be found by leaving the M54 at

Junction 5 and taking the Rampart Way (A5T) to the Hollinswood Interchange. The Town Park can be entered from Stirchley Grange, and also from a car park near the junction of Randlay and Stirchley Avenues ½ mile (0.8 km) further north, or from the Town Centre.

The Stirchley Interchange is also the point to leave Queensway (A442) for Holmer Lake to the south-west and Nedge Hill to the north-east. Donnington Wood is in the north-east of Telford and can be reached from the Wombridge Interchange of Queensway via Wrockwardine Wood Way (B4373). The two Priorslee waters lie to the north of the M56 between Junctions 4 and 5 and can be reached from either.

The Ironbridge Gorge forms the southern boundary of Telford, and there are car parks and picnic sites on the B4373 from Ironbridge and on the unclassified road following the north bank of the river to Coalport. Lightmoor can be approached by taking the A4169 north from Ironbridge and taking the right turn into Cherry Tree Hill road at the foot of Jiggers Bank.

## Calendar

*Resident:* Little and Great Crested Grebes, Heron, Mute Swan, Canada Goose, Mallard, Tufted Duck, Sparrowhawk, Kestrel, both partridges, Woodcock, Little and Tawny Owls, Kingfisher, all three woodpeckers, Grey Wagtail, six tits, resident thrushes, Nuthatch, Treecreeper, resident finches, Yellowhammer, Reed Bunting.

*April–June:* Passage waders including Ringed and Little Ringed Plovers, Oystercatcher (scarce), Redshank. Terns. Summer migrants include Turtle Dove, Cuckoo, hirundines, Tree Pipit, Sedge Warbler, scrub warblers, leaf warblers, flycatchers.

*July–September:* Residents and summer migrants still breeding. Return passage of waders, including Greenshank and Green Sandpiper. Terns.

*October–March:* Cormorant, Wigeon, Teal, Shoveler, Pochard, Golden-eye, occasionally scarcer ducks, winter thrushes, Siskin, Redpoll, Crossbill (scarce).

## BENTHALL EDGE WOOD

### Habitat

This wood is on the southern slopes of the Ironbridge Gorge, and staddles a complicated geological area made up largely of limestones and coal measures. The variety of the underlying rocks leads to a great diversity of plant life, which includes several species of orchids. There are many traces of mineral extraction — here was the cradle of the Industrial Revolution — but most are now clothed in deciduous woodland rich in wildlife. Badger, fox and fallow deer are amongst the mammals that may be encountered here. Ash, beech, birch, wych elm, hazel, holly, oak and the much maligned sycamore all flourish, with at least a dozen other tree species present. Management policy is to remove some sycamore to encourage native trees such as ash and elm.

The wood grows on a very steep slope, and is well served by footpaths. A nature trail of 1¾ miles (2.8 km) has been constructed to show off the more interesting features, but for those requiring a less strenuous walk, the old railway track can be followed until reaching the power station boundary. This route gives views onto the River Severn in places, and regularly provides sightings of Mallards, Kingfishers and Grey Wagtails.

## Species

Observers walking along the old railway track, especially early in the day, may see considerable birdlife on the river. Many species use the Severn as a flight route, and Cormorants, Grey Herons, Mute Swans, Mallards, gulls, and migrating waders and terns pass through. Low across the water the jerky flight of Common Sandpipers may be seen, the brilliant flash of a Kingfisher, or the slender, long-tailed shapes of Grey and Pied Wagtails as they shoot along in characteristic fashion. An unexpected sighting in 1981 was an Osprey, seen fishing the river a little downstream from the Iron Bridge.

The woodland abounds in birds in spring and summer. Tree-holes are plentiful and consequently the species which nest in them are well represented, including Stock Doves, Tawny Owls, Redstarts, Nuthatches, Treecreepers, Jackdaws and Starlings. Marsh, Coal, Blue and Great Tits also seek ready made sites, whilst Willow Tits find suitable timber in which to excavate their own holes, as do all three woodpeckers. Pied Flycatchers are further encouraged by the provision of nest-boxes. Warblers are a feature of the wood, advertising their presence by loud song, yet difficult to locate once the trees are in full leaf. Garden Warblers, Blackcaps, Wood Warblers, Chiffchaffs and Willow Warblers are all numerous. Many other common woodland songsters are present, and with so rich a population of small passerines, Cuckoos are on hand to practice their parasitic habits, and Sparrow-hawks find a satisfying supply of food. Buzzards are seen above the wood on rare occasions, and evening visits during spring should ensure sightings of roding Woodcocks.

The upper boundary of the wood meets agricultural land with small fields and hedgerows. It is worth exploring here to locate species of the woodland edge, where Kestrels hunt for small mammals and Jays are often seen commuting between fields and wood. Spotted Flycatchers, Long-tailed Tits and Mistle Thrushes breed, with Fieldfares and Redwings feeding on the autumn berries.

## Timing

The Ironbridge Gorge has become a 'Mecca' for tourists in recent years, but most of them are absorbed by the attractions of the town, especially by the excellent museums. Away from the peak holiday period, and weekends during fine settled spells, Benthall Edge Wood is surprisingly free of disturbance. April to June is the best time to visit, preferably before the trees are in full leaf.

As usual with woodlands, an early morning visit will bring the greatest reward, and as this is not an area for soaring raptors the mid-day period is best avoided. Evenings can also be very enjoyable, with calling Tawny Owls, roding Woodcocks, and perhaps even the far-carrying song of a Nightingale from the coppices downstream.

## Access

Ironbridge lies 4¾ miles (7.6 km) south of the A5(T), almost due south of Wellington, from which it can be approached along the A442 before turning right onto the A4169 after 1¾ miles (2.8 km). This road enters the Gorge after turning left at the foot of Coalbrookdale.

There are several large car parks available, but the best one for Benthall Edge Wood is situated at the southern end of the famous Iron Bridge which now takes pedestrians only. The best starting point is along the old railway track. There is a picnic area here, and the start of the nature trail. If intending to ascend to the top of the Edge, birdwatchers should wear strong footwear and be prepared for muddy and slippery paths, which may take a considerable period of time to dry out after heavy rain.

## Calendar

*Resident:* Sparrowhawk, Woodcock, Stock Dove, Tawny Owl, all three woodpeckers, common woodland song birds, Long-tailed and five other tits, Nuthatch, Treecreeper, Jay.

*April–June:* Woodcock roding. Cuckoo and other summer migrants arrive from mid-April, including Redstart, Garden Warbler, Blackcap, Wood Warbler, Spotted and Pied Flycatchers.

*July–September:* Residents and summer migrants breeding. Mixed flocks of juvenile tits.

*October–March:* Fieldfare and Redwing on woodland edge. Tawny Owls noisily dispute territories. Resident birds take up territory from late February. Chiffchaff arrives late March.

## CRESSAGE TO BUILDWAS

### Habitat

The Severn Valley is wide and flat in this area, resulting in spectacular meanders and awesome flooding. On the former, the greater speed of the current at the outside of the bends erodes the banks to form long stretches of vertical cliffs suitable for hole-nesting Kingfishers and Sand Martins. On the inside of these meanders the slower current deposits mud, sand and pebbles to create beaches used by many species for resting, bathing, preening, feeding and breeding. Winter levels submerge these beaches and much of the bankside vegetation. Unseasonal rises in the river can cause serious damage to riverside breeders, especially grebes (which anchor their nests to trailing willow branches) and Moorhens. Winter flooding creates a vast lake in the valley, and even after the water has returned between the banks there are pools remaining in hollows for several days or even weeks. In long periods of freezing, with all lakes iced over, many birds take to the river until milder weather returns.

Several small eyots, some vegetated, add interest to this section of the Severn, but all except one disappear when the river flows at winter level. The range of vegetation is limited by the constant ravages of the water, with sedges and course-grasses the norm, whilst willow is the dominant tree. Oaks, widely spaced, make ideal drying posts and roosting sites for Cormorants. The fields bordering the river are a mixture of pasture and arable, but all are well drained so that the damp meadow situations enjoyed by Redshanks and other waders do not occur.

A walk upstream from Cressage Bridge is also recommended. This length of river has extensive pebble beaches, a large area of thorn scrub and an interesting hillside wood.

## Species

Regular breeding birds in the riverside vegetation are Moorhens, Sedge Warblers and Reed Buntings. The willows attract many species, and bands of tits, usually Long-tailed, Blue and Great Tits, move from bush to bush along the river. The willows also make a good roosting site, and are used by Magpies, Fieldfares, Redwings and other thrushes. Lesser Whitethroats skulk in them at the end of most summers on passage. Kingfishers and Sand Martins tunnel nest-holes in the vertical cliffs.

The pebble beaches and eyots exposed at low water periods are favourite haunts of Little Ringed Plovers and Dunlin on spring passage, with the elegant Greenshanks and Green Sandpipers prominent during the return movement. Common Sandpipers, flicking low above the river surface, appear in numbers at both times. Grey Wagtails and Dippers feed in these areas, although they breed on smaller tributary streams, whilst many species including wildfowl, Lapwings, Curlews and gulls gather on them to bathe and preen.

In the fields and hedgerows a wide range of species can be found, the area being particularly good for Tree Sparrows, not a common bird in Shropshire. Stubble, something of a rarity these days, is often a 'Mecca' for finch and bunting flocks from autumn, with Chaffinches, Greenfinches, Goldfinches, Linnets and Yellowhammers present in large numbers. Bramblings and Redpolls are more likely at the end of winter. Such an abundance of prey, along with voles and field mice, attract predators, and Sparrowhawks, Kestrels, Little Owls and Tawny Owls are regular here, with occasional Merlins and Peregrines putting in an appearance during winter months. Sadly, Barn Owls are more conspicuous by their absence these days, although irregular sightings still occur, as do those of Short-eared Owls. Skylarks and both partridges are residents in these fields, and other breeding birds are Lapwings, Curlews and Yellow Wagtails. The meadows to the east of Leighton Hall are an important wintering ground for wildfowl, and up to 1,000 Canada Geese can be counted there in December, with very small numbers of Pink-footed, White-fronted and Brent Geese dotted amongst them, and feral Greylag, Snow and Barnacle Geese adding interest. One or two Bar-headed Geese appear in most winters, escapees from wildfowl collections. Wigeon flock here each winter, building up to 400 or so at the turn of the year.

Weather conditions, which dictate the state of the river, bring different species to the area accordingly. A long period of winter rain will cause severe flooding, and birds attracted at these times are Bewick's Swans, Whooper Swans, Pintail and Shoveler, the latter remaining on flood pools after the water has receded. Mats of flood debris, floating on the river surface, but trapped between the trunks and branches of willows, obviously provide a useful source of food, as many small birds, including Siskins, feed on them, unlikely though it may seem. Prolonged freezing will increase the numbers of both the common grebes, Pochard, Tufted Duck, Goldeneye, Goosander and Coot. A clear sunny day during a freeze-up, with the river reflecting a blue sky, and the smartly pied drakes of Tufted Duck, Goldeneye and Goosander diving and resurfacing between ice-floes can be quite magical, especially if there is a carpet of snow to add to the Arctic effect. A falling river level in these conditions will expose a narrow strip of unfrozen banking in a rock-hard countryside, which Snipe and Song Thrush are always quick to exploit.

Other birds which are commonly seen in the area are Cormorants, outside the breeding season — with sometimes a score of birds sitting in a riverside oak — Grey Herons, Mute Swans and Mallard. All the commoner gulls are present, especially on flood pools, with Black-headed and Lesser Black-backed Gulls the most numerous, and the occasional Kittiwake passing through in late winter.

Mink are a recent addition to the fauna of the river, and are reputed to be a threat to native wildlife, especially Moorhens and other riverside nesting birds.

On the upstream side of Cressage Bridge, beneath which Kingfishers regularly fly, the pebble beaches are good for the wader species already mentioned, and Redshanks appear each spring. They were annual breeders in a field which has recently been improved for pasture. The thorn scrub provides shelter, food and nest-sites for thrushes, warblers, tits, finches and buntings. Newly arrived Fieldfares and Redwings feed on them in hundreds in autumn. The hillside wood holds a good selection of birds, including Sparrowhawks, all three woodpeckers, Nuthatches and Treecreepers.

## Timing

There is something of interest throughout the year, with perhaps the leanest spell falling between the departure of summer migrants and the arrival of wintering birds. The winter months produce good birdwatching, especially when pools and lakes in the county are frozen, or the river is in full flood, but be prepared for footpaths being inaccessible, and footbridges being damaged or destroyed. Dry spells are also very good, with maximum exposure of beaches and eyots. The worst river condition is when the water runs bank-full but is not flooding onto the fields.

There is some disturbance from anglers, although the majority, once in position, have little effect on birdwatching. There is also a small amount of shooting in season, but this is insignificant. In summer months canoeing activity may cause temporary disquiet, but parties usually pass quickly through. By far the greatest amount of disturbance on the Shropshire Severn in recent years has been caused by the proliferation of Raft Races, which are noisy social affairs, and the river is best given a miss on these occasions. The field at Buildwas is sometimes used as a starting point.

Early morning is the best time of day to visit at any time of the year. Evenings can produce spectacular flights of wildfowl, especially in October when hundreds of Canada Geese fly along the Severn Valley, heading west from the Leighton meadows. From October to March evening flights of Wigeon are a delight to see and hear.

## Access

There is parking space at Cressage Bridge on the spur of the B4380 that runs into the village from the north. Follow the footpath upstream as far as the second river bend. It may be necessary to take the route around the outside of the small wood due to recent severe bank erosion, and it is advisable always to go this way when the river is flooded. On the other side of the road a path leaves the field gate and follows the hedgerow towards Leighton, touching the river for a distance from near Eye Farm.

The river is also accessible in the east of the area. Park on the old road beside Buildwas Church on the B4380 at the west of the village. The Ordnance Survey map shows a path running through fields, but these are invariably ploughed and the line of the path not maintained. Take the anglers' path alongside the river from the stile 220 yds (200 m) east of the church. Footbridges cross the small tributary streams and the path can be followed for about 1¼ miles (2 km).

The inaccessible private areas can be viewed from the lay-bys overlooking Leighton Park on the B4380, 1 mile (1.6 km) west of Buildwas Church. Grandstand views are obtained from here, but it is necessary to use a telescope for satisfactory viewing.

## Calendar

*Resident:* Grey Heron, Mute Swan, Canada Goose, Mallard, Sparrowhawk, Kestrel, Red-legged and Grey Partridge, Moorhen, Lapwing, Little and Tawny Owls, Kingfisher, Skylark, common resident thrushes, tits and finches, Tree Sparrow, Yellowhammer and Reed Bunting.

*April–June:* Passage Little Ringed Plover, Dunlin, Redshank and Common Sandpiper. Breeding Curlew, Sand Martin, Yellow Wagtail and Sedge Warbler. Resident birds breeding.

*July–September:* Greenshank, Green and Common Sandpiper on passage. Lesser Whitethroat (passage) and tit flocks in willows. Canada Goose and Lapwing form post-breeding flocks.

*October–March:* Little and Great Crested Grebes return to river. Cormorants return from coast. Winter wildfowl arrive, Bewick's Swan, Whooper Swan (rare); Pink-footed (scarce), White-fronted and Brent Goose (rare); Greylag, Snow and Barnacle Goose (all feral); Wigeon, Pintail, Shoveler, Pochard, Tufted Duck, Goldeneye and Goosander. Gulls on floods. Kittiwake (scarce) in March. Grey Wagtail and Dipper move down from smaller streams. Fieldfare and Redwing. Large finch flocks include smaller numbers of Brambling, Siskin and Redpoll. Very occasional Merlin and Peregrine. Sand Martins arrive at end of period.

## THE WREKIN AND THE ERCALL

### Habitat

Aligned from north-east to south-west, these hills of ancient volcanic lava lie north of the River Severn. The Wrekin is 2 miles (3.2 km) and The Ercall almost 1 mile (1.6 km) in length. Although not particularly high — the former reaches 1,334 ft (407 m) and the latter 871 ft (265 m) — the range is made to look impressive by the flatness of the Shropshire Plain from which it rises. Both hills are thickly wooded, but the Wrekin has some open grassland along the summit ridge and a few outcrops of rock. The prominence of the hill made it a natural choice for the site of an Iron Age fort and later as a beacon.

181

Much of The Wrekin woodland is modern forestry plantations, especially along the south-eastern slopes and at the southern end, but some areas of broad-leaved woods remain, including oak, with an extensive beech wood along the north-west slopes towards the northern end. There are recently felled and replanted areas, suitable for Grasshopper Warblers, and others should be created periodically as the conifers mature.

The Ercall is clothed mainly with broad-leaved trees, but is scarred in part by a working quarry in the south. It provides pleasant birdwatching, but does not afford the spectacular views for which The Wrekin is renowned, nor is there as much variety in the habitat.

A small pool lies in the angle between two unclassified roads, one of which separates the hills. This is the old Wrekin Reservoir. On the other two sides, undisturbed rough grass and scrub, together with a bordering stream and a number of mature trees, creates a habitat which is often rich in birdlife, although only a limited number of species use the pool.

## Species

A quick look at the pool may reveal Little and Great Crested Grebes, with Mallard, Pochard and Tufted Duck the likeliest wildfowl. Moorhens and Coots both breed. The surrounding vegetation holds scrub loving species such as warblers in summer, and provides a good feeding ground for finches in winter, including Siskins and Redpolls, but also Greenfinches, Goldfinches and Linnets.

The coniferous plantations of The Wrekin shelter Goldcrests, Coal Tits and occasional Crossbills, whilst the newly replanted areas are favoured by Grasshopper Warblers, and where isolated birches have been left unfelled Tree Pipits breed, followed by Garden Warblers, Blackcaps and Linnets as the scrub situation develops. Whinchats may also be seen here, and there was a report in 1983 of Nightjars in the area (but sadly this was followed by the finding of a dead road victim).

The usual broad-leaved woodland species are found on both hills. Redstarts, Wood Warblers and Pied Flycatchers all breed. Most common woodland songbirds are present, with spring days resounding with melodies of Wrens, Robins, Blackbirds, Song Thrushes, Garden Warblers and Blackcaps. Turtle Doves croon, and Cuckoos become monotonous after the first few exciting calls of the early arrivals. Hole-nesting species include five tits, Nuthatches, Treecreepers, all three woodpeckers, Stock Doves and Jackdaws. The main predators are Sparrowhawks and Tawny Owls, although Jays will also be a threat to eggs and nestlings of many passerines. Spring and summer evenings are enlivened by roding Woodcocks, and bats are numerous in clearings and along woodland edges. Siskins and Redpolls, seeking birch, spruce and larch seeds are present during most winters, as are Bramblings, feeding on fallen beechmast with Chaffinches.

Along the open ridge, and around the summit rocks, Kestrels hunt and a few Meadow Pipits may be seen. Swifts, Swallows and House Martins feed across and above the rock faces on which Carrion Crows love to perch, and about which Jackdaws demonstrate their considerable flying skills. Other species enjoying the more open areas and hedgerows of neighbouring fields are Yellowhammers, Mistle Thrushes and the winter visiting Fieldfares and Redwings. Neither Buzzards nor Ravens breed here, but both are occasionally seen on The Wrekin.

# Timing

The Wrekin is a top tourist attraction for hordes of visitors to Shropshire. Most of these tackle the hill from the road at the northern end, following the well-worn path to the summit. Summer weekends are the worst affected, but fine weather at any time of the year can bring visitors in some numbers. Joggers, orienteers, pony riders and ramblers are all attracted to the area, but most disturbance can be avoided by an early morning visit or by taking some of the less used paths.

There is a Scout Camp on the south-east side. A thrash around the hill during dusk and darkness seems to be some sort of tradition with this movement, so if the tranquillity of the evening is your preference, with roding Woodcocks, hooting Tawny Owls, bats on the wing, and perhaps a prowling fox, you must avoid Fridays and Saturdays.

A small military firing range on the north-west side may cause temporary inconvenience, but this is not a major disturbance to the hills generally and poses no threat to public safety. (Warning flags are flown and sentries posted when the range is in use.)

By comparison The Ercall is reasonably quiet, although it is popular with local people, especially family parties.

Spring and early summer are the best times to visit, but there is something of interest at most times of the year. August and September are probably the leanest months.

# Access

This area is well served by footpaths. The road on the south-west side of the reservoir has a wide verge suitable for parking and can be used for both hills. It is reached by taking one of the unclassified roads south from the A5(T), or by meandering through the network of lanes west from the A4169 or north from the B4380.

There is limited roadside parking at the south-west end of The Wrekin, but the paths through the plantations at this end are very muddy in wet periods and footwear should be chosen to suit. With patience and a good navigator the lanes listed above will get you to this spot.

# Calendar

*Resident:* Sparrowhawk, Kestrel, Woodcock, Tawny Owl, three woodpeckers, Nuthatch, Treecreeper, Goldcrest, Jay and other common crows, resident thrushes and finches, tits.

*April–June:* Summer visitors arrive, mainly mid-April, including Cuckoo, Tree Pipit, Redstart, Pied Flycatcher, Wood Warbler, scrub warblers; but early May for Swift, Turtle Dove and Spotted Flycatcher. Roding Woodcock.

*July–September:* Residents and summer visitors still breeding. Fledglings on wing. Flocks of mixed juvenile tits in woods.

*October–March:* Grebes and wildfowl on reservoir (small numbers). Occasional Buzzard and Raven. Fieldfare and Redwing from mid-October. Winter finch flocks include Brambling, Siskin and Redpoll. Crossbill (scarce). Chiffchaff arrives end of March.

# LONG MYND, STIPERSTONES
# & THE STRETTON HILLS

## Habitat

This whole area is one of outstanding scenic beauty, and on clear days any high point in the hills offers staggering views in all directions. The three groups are very different in character, from the brooding rocky ridge of the Stiperstones, across the sprawling plateau of the Long Mynd, to the hog-backed Stretton Hills.

The Long Mynd runs north-north-east from Plowden before merging into the patchwork of agricultural land 8 miles (12.8 km) to the north. The greatest width is 4 miles (6.4 km) between Church Stretton and Ratlinghope. This upland is renowned for its geological interest, the pre-Cambrian rocks forming a plateau rising to an average height of 1,500 ft (460 m), with a highest point of 1,695 ft (517 m) at Pole Bank. The area is dotted with ancient tumuli and dykes, with Bodbury Ring, an Iron Age hill fort, overlooking the Cardingmill Valley.

The habitat is varied, but dominated by heather moorland, most of which is owned by the National Trust (NT). The heather is managed as a grouse moor, being cut periodically to produce new growth without encouraging the spread of bracken. Bilberries also grow in profusion, and attract both human and wildlife to harvest the crop. Bracken does dominate in some areas, and seems much loved by Whinchats. Gorse flourishes locally, and apart from enlivening the scene with bright golden-yellow flowers, also provides song posts for Yellowhammers and occasional Stonechats. Small pools and boggy patches lie hidden between acres of heather, insectivorous round-leaved sundew and common butterwort being found in the latter. Parts of the Long Mynd have been converted to improved grassland, and the whole area is grazed by sheep.

Pure streams flow from springs surrounded by sphagnum moss, to cascade into spectacular valleys cut deeply into the eastern facing slopes. Known as 'batches', their steep sides are generously dotted with scrub hawthorn and rowan, and small rocky outcrops push through the heather in most valleys. The western facing slopes are steep, producing updraughts suitable for gliding, hang gliding and raptors, especially Buzzards.

Nestling beneath the slopes of the Long Mynd, on the outskirts of Church Stretton, is Old Rectory Wood. Owned by Shropshire County Council (SCC), the wood contains a good mixture of trees, mainly hardwoods, including many beech, with oak, chestnut, cherry and Scots pine amongst others. A nature trail has been constructed to cover the most interesting aspects of the wood, and it is an excellent site for all woodland species.

Less attractive, both ornithologically and aesthetically, is the blanket coniferous plantation towards the southern end of the Long Mynd. Even so, it is worth a look for its resident Goldcrests and Coal Tits. It also provides nesting sites and shelter for birds feeding on adjacent hillsides, especially Jays, which regularly commute between wood and hill. Sparrowhawks and Goshawks (rare) are also encountered here.

The barren, rocky ridge of the Stiperstones is one of the wildest areas in Shropshire. To walk the hill on one of the not infrequent days of swirling mist can be an unnerving experience. To be caught on the ridge during a summer thunderstorm can be downright frightening, and it is easy to see why the area is steeped in legend. Lying parallel to the Long Mynd, some 4 miles (6.4 km) to the west, it was once a continuous quartzite ridge before excessive frost shatter in the last Ice Age reduced it to the line of jagged tors we see today. The most impressive stretch is the 1¼ miles (2 km) between Cranberry Rock and Scattered Rocks, north of the Bridges to Shelve road. The whole length here is boulder strewn, and strong footwear is recommended. Heather and bilberry make up the dominant vegetation, with some cowberry and the scarce crowberry. Scrub hawthorns and rowans are invading the lower slopes. Extensive planting of conifers at the southern end and on the eastern side is a comparatively recent development, which brought Nightjars to the area for a few years, although they now appear to have left. The Stiperstones lacks the streams of the Long Mynd, but there are interesting wooded valleys in the north-west at Mytton Dingle and Perkins Beach. The area surrounding the hill is made up of small hill farms and the remains of the former lead mining industry. The Nature Conservancy Council (NCC) declared a large portion of the Stiperstones to be a National Nature Reserve (NNR) in 1982.

Also parallel to the Long Mynd, and forming the eastern slopes of the Vale of Stretton, the hog-backed Stretton Hills lie in line ahead, aligning on their more northern cousin, the Wrekin. The pre-Cambrian rocks, some of the oldest in Britain, form narrow, steep sided hills, clothed in acid grasslands, with occasional rocky outcrops making ideal lookout perches for Ravens and Kestrels. Some patches of oak woodland are to be found, the most extensive being on Helmeth Hill and Ragleth Hill. The dominating hills are Ragleth at 1,300 ft (396 m), Caer Caradoc at 1,506 ft (459 m) and the Lawley at 1,236 ft (377 m), but the outlying hills to the east are worthy of exploration. Willstone Hill, Middle Hill and Hope Bowdler Hill form a close group, more flat-topped than their neighbours, but rising to similar heights.

## Species

These hills are excellent raptor haunts. Hovering Kestrels may be seen anywhere, as may soaring Buzzards; and Sparrowhawks hunt the entire area, even skimming the heather of the open moors. All three species breed, and families of Kestrels and Buzzards are frequently seen in late summer, lined up with military precision in the updraughts of western facing slopes. Hen Harriers are passage birds, seen quartering the open moorland of the Long Mynd and the Stiperstones. Seasonal visitors are Hobbies, summer migrants now seen annually, and Rough-legged Buzzards, which occur in some winters. The former may be seen hunting over the Long Mynd, occasionally as a pair, for fox moths and northern eggar moths, whilst Rough-legged Buzzards are usually seen in the north-west of the Long Mynd, and the marginal hill land between there and the Stiperstones. Peregrines, being great wanderers, may appear anywhere at any time of the year, often as immature birds, and Goshawks have been seen more frequently in recent years, particularly in spring. The high population of Woodpigeons should ensure that both species find local food supplies to their liking. The delightful little Merlin is seen hunting the heather moors in most years, when Meadow Pipit numbers are high, and occasional winter birds are seen in bordering lowland pastures. Red Kites also wander into the hills, usually in winter, and if the Welsh population does eventually manage to increase significantly, annual sightings should become the norm.

A number of other species are common throughout the area. Meadow Pipits are very numerous and play host to Cuckoos, but Tree Pipits are more local, occurring along woodland edges and on thorn covered slopes. Up to 500 Swifts have been estimated feeding above the Long Mynd, but along with Swallows and House Martins they feed over all the hills, particularly when ants are on the wing. Curlews breed widely, and smaller numbers of Snipe perform evening drumming flights over damp patches. The ubiquitous Carrion Crow is rarely out of sight or earshot, and large flocks of Rooks and Jackdaws feed in the hills. Ravens breed here, their croaking calls and thrilling flight being symbolic of wild hill country, and more than compensating for the lack of physical beauty. Magpies and Jays, handsome but noisy, are commonplace, and Choughs have appeared on very rare occasions after westerly gales. Mistle Thrushes, lovers of soft fruits, gather in large numbers to feed on bilberries from early summer. Other passerines common to all areas are Wheatears, especially around rocky outcrops with short turf, Linnets and Yellowhammers, the latter two species associating with scrub hawthorn and gorse.

Red Grouse breed in some numbers on the heather moorland of the Long Mynd and the Stiperstones, and Short-eared Owls are vagrants. Stonechats, though scarce, usually breed in both places, but the numerous Whinchats tend to move onto the valley slopes, and post-breeding families often associate with bracken, perching conspicuously on the taller fronds.

The small pools on the Long Mynd provide breeding sites for Mallard, Teal and occasional Moorhens. Green Sandpipers also use them, usually for a very short stay, pausing only to bathe and preen before continuing their journey. Wet areas, habitat of the soft rush, are sought out by breeding Reed Buntings in summer, and are frequented by solitary Jack Snipe in late winter. Lapwings feed on the improved pastures, as do Skylarks, and other Long Mynd waders are Golden Plovers, seen on rare occasions in spring, and the charming trips of Dotterels, seen almost annually on spring and autumn passage, usually near the Gliding Field.

The Long Mynd valleys are excellent for wildlife when the harshness of winter has passed. Grey Wagtails, Pied Wagtails and Dippers breed along the streams, and Herons visit them for small fish. The latter also visit the moorland pools for frogs. The scrub on the valley slopes is often teeming with small passerines such as Great Tits, Blue Tits, Chaffinches, Willow Warblers and Robins. Dunnocks and Wrens breed almost to the edge of the open moor, and all the major valleys hold Ring Ouzels and Redstarts in small numbers. The latter often favour partially dead windblown thorns, lying grounded with bracken growing into the branches. Green Woodpeckers search the slopes for ants, and Red-legged Partridges call from rock outcrops, looking far more impressive than their lowland brethren.

Old Rectory Wood, and other patches of hardwood found at the foot of most valleys contain Tawny Owls, Great and Lesser Spotted Woodpeckers, Wood Warblers, Nuthatches, Treecreepers and all the usual woodland passerines. The coniferous plantation shelters Goldcrests and tits, with a larger variety of passerines along the perimeter, including Siskins, which are found on the lower boundary in Minton Batch.

*Ring Ouzels breed in the valleys of the Long Mynd*

The Long Mynd is always capable of springing a surprise, and recent years have seen a Ringed Plover, Black-tailed Godwits, Wood Sandpiper and Sandwich Terns overflying on passage, and a flock of disorientated Canada Geese flying at heather-top height through thick fog.

Although less extensive in range and more restricted in habitat than its neighbour, the Stiperstones offers good birdwatching. The abandoned fields and hedgerows surrounding the ruins of former mineworkers' cottages in the north are good for Redstarts, Pied Flycatchers, tits and other small passerines. The coniferous plantation north of the Bridges to Shelve road contains large flocks of small birds in autumn, especially Goldcrests and Coal Tits, with birds commuting between here and the oak wood to the east. The northern end of the plantation is more open, containing mature Scots pine, and the three woodpeckers are found here, with Woodcocks roding on spring evenings. The scrubby hillside between the plantation and the car parking area is good for pipits, Whinchats, Linnets and Redpolls. Ring Ouzels are not seen on the Stiperstones every year, but have bred recently in Perkins Beach, an area much frequented by Jays.

Apart from species already mentioned, birdlife on the tops of the Stretton Hills is rather sparse, but the fringe country can be very good. Redstarts in old hedgerow timber, Nuthatches, Treecreepers, the three woodpeckers and Tawny Owls are all typical birds of the area, and winter thrushes join resident thrushes to feed on hawthorn and rowan berries in autumn. The patch of country containing Willstone, Middle and Hope Bowdler Hills is rather neglected by birdwatchers and could hold surprises for the visitor.

## Timing

The Shropshire Hills are a tourist area and attract great numbers of visitors from Easter through to September. At first, and towards the end, it is mainly weekends when the crowds arrive, but July and August can be chronic each day. The Long Mynd suffers most, but it is a vast area and most people do not stray far from their vehicles. It is, however, a playground for outdoor activities of many kinds (including birdwatching!) and for educational visits, which may involve more than 100 children spending school hours in one of the valleys. The Cardingmill Valley and Ashes Hollow are usually worst hit. The Stiperstones take less visitors, but the area is more compact and cannot absorb large numbers so effectively. Again, July and August are the worst months, with many people coming specifically to pick bilberries. On the Stretton Hills, ridge walking is a popular pastime, but not particularly disturbing to birdlife.

One sure way to avoid disturbance is to visit early in the day. In spring and early summer the hills are alive with bird movement and song. Return at noon and the difference is unbelievable. A word of warning though. Temperatures can be below freezing at dawn even in May, and showers of snow or hail are not unusual in that month. Rain can be sudden and heavy, even when the lowlands are fine, so wrap up well and carry spare clothing if you are walking any distance. Evening visits can be very profitable too, and the tranquillity of dusk, disturbed only by the calls of Red Grouse and Curlew, the drumming of Snipe and the sudden appearance of a late hunting Hobby, is something very special.

Winter birdwatching is usually very poor, especially at high altitudes, although the occasional Raven or raptor may make it worthwhile. Most birdlife at this time is confined to the lower valleys, woods and marginal land.

## Access

There is open access to those parts of the Long Mynd owned by the NT, and the area is well served by roads. The most popular route is from Church Stretton in the east, but alternative roads may be taken from Bridges in the north-west, and from Plowden via Asterton in the south. The northern end can be reached from Leebotwood on the A49(T). Old Rectory Wood is within easy walking distance of Church Stretton.

The most popular access point for the Stiperstones is the car parking area on the Bridges to Shelve road, below Cranberry Rock. Other good starting points are at Lordshill, near Snailbeach, in the north, and Stiperstones village in the north-west. The area is well served by rights of way, and there is open access to the Stiperstones NNR.

The Stretton Hills are also well served by footpaths. The Lawley is best approached from the minor road passing the north-east end. Hoare Edge, on the opposite side of the valley, can also be gained from this road. Caer Caradoc and Helmeth Hill can be approached from Willstone to the east (the narrow lane from the farms is unsuitable for cars), or from the B4371 beneath the Gaer Stone, 1 mile (1.6 km) east of Church Stretton.

This point also gives access to Willstone, Middle and Hope Bowdler Hills, which can also be walked from paths leaving the same road near Hope Bowdler village or ½ mile (0.8 km) further east.

Ragleth Hill can be reached via Clive Avenue in Church Stretton through woodland in the north-east, or from the A49(T) at the south-west end.

## Calendar

*Resident:* Sparrowhawk, Buzzard, Kestrel, Red Grouse, Red-legged Partridge, Woodpigeon, Stock Dove, Woodcock, Tawny Owl, the three woodpeckers, Dipper, Wren, Dunnock, resident thrushes, Goldcrest, six tits, Nuthatch, Treecreeper, common crows, Raven, resident finches, Yellowhammer.

*April–June:* Breeding Mallard, Teal, Lapwing, Snipe, Curlew, Skylark, Stonechat and Reed Bunting on moors; Grey Wagtail, Pied Wagtail and Dipper in streams. Hen Harrier and Dotterel on passage. Summer migrants from mid-April include Cuckoo, Tree Pipit, Redstart, Whinchat, Wheatear, Ring Ouzel, Wood Warbler and Pied Flycatcher. Occasional Herons. Hunting Hobby and other raptors.

*July–September:* Hen Harrier and Dotterel on passage. Family parties of hunting raptors. Soaring Buzzard families. Juvenile Cuckoos. Large gatherings of Swifts until mid-August. Mistle Thrushes gather for bilberries, and passage Ring Ouzels feed on rowan.

*October–March:* Rare visits by Red Kite and Rough-legged Buzzard. Peregrine, Jack Snipe, Short-eared Owl (all scarce). Flocks of Fieldfare and Redwing. Finch flocks include Siskin and Redpoll. Resident birds move back to the high ground at end of the period.

This is a flat agricultural area, dotted with small woods and pools, although the majority of the latter are not noted as bird-rich sites, and many are on private property. Shrewsbury, the county town, is situated on the River Severn, which flows through the countryside in a series of sometimes spectacular meanders. The town centre is almost completely surrounded by one of these huge loops, and a walk along the riverside path from the English Bridge on the east to the Welsh Bridge on the west can produce interesting species. Wildfowl escapees such as Mandarins, Carolina Wood Ducks and Red-crested Pochards frequently put in an appearance. Mute Swans, Mallards, Tufted Ducks and Moorhens are regulars, with Little Grebes occasionally seen. Kingfishers are not unusual, and one fishes regularly from bushes near the Welsh Bridge. Some winters ago a Shag spent several weeks on the river, often resting on the English Bridge, much to the delight of local birdwatchers, and more recently a Common Scoter visited the town.

## MONKMOOR

### Habitat

In the Monkmoor area the river is at the extreme point of a major meander. Shrewsbury Sewage Farm is situated within this meander, and was formerly the foremost wader site in the county. There is no public access to the sewage farm, but a pool which can be viewed from the parking area, at the end of the road from Monkmoor, often holds a wide range of species. The river banks are generally steep, providing suitable nesting cliffs for Kingfishers and Sand Martins, and the narrow riverside meadows on the inside bend are left as pasture for grazing cattle. Opposite, the bank is partially lined with willows and the adjoining fields are arable, with hedgerow removal to the south of Uffington creating what must be one of the largest fields in Shropshire. This reach of the Severn also contains a well-vegetated eyot which provides shelter and breeding sites for a number of birds, especially Mallards, Sedge Warblers and Reed Buntings. A small tree-lined stream entering on the northern side of the river is often a focal point for birds feeding in the area.

### Species

Little Grebes and Great Crested Grebes may occur on the pool throughout most of the year, and are seen on the river during winter months. Other birds associated with both standing and running water are Mute Swans, Mallards and Moorhens. Coots move to the river during spells of prolonged freezing. Sedge Warblers and Reed Buntings breed regularly in both habitats.

Birdlife on the river is seasonal. Cormorants fish its waters outside the breeding season, Herons visit throughout the year, except in times of high water level, and Goosanders are encountered in small numbers during winter months. Passage periods bring sightings of waders and terns. Of the waders, Oystercatchers and Common Sandpipers are seen each spring, with the latter returning early in the autumn, followed by Green Sandpipers, Greenshanks and the occasional Curlew Sandpiper. Terns are usually Common and Arctic Terns, but a Little Tern graced this stretch of river in 1978. Kingfishers breed, but may be scarce following harsh winters. Sand Martins, low in numbers during recent seasons due to drought conditions in their winter quarters, appear to have survived the crisis and can be seen feeding above the river from the final week of March. Grey Wagtails are occasionally seen, whilst Sparrowhawks and Kestrels hunt riverside fields and hedgerows.

Tufted Ducks are resident on the pool, and are joined in winter by Teal, Shoveler and Pochard. All may be forced to the river in freezing spells. Ruddy Duck visit the pool in spring and autumn. Spring is also the best time to see the delightful Yellow Wagtails, with males resplendent in breeding finery, and a very rare Red-rumped Swallow put in an appearance in 1978. Hobbies have been recorded here too.

## Timing

For passage waders and terns April, May and August to early September are the best periods. Post-breeding Cormorants start arriving in September, but it is usually mid-November before winter wildfowl appear in numbers.

The spot is well-liked by anglers, and attracts a few local walkers, but disturbance is never great. Early morning is the best time of day to visit, particularly in spring and summer. For most species low water levels are an advantage, especially waders, Herons and wagtails, and conversely a bank-full river of rushing muddy water will produce few birds. A visit during or within two or three days of heavy rain will be sure to disappoint.

## Access

From the large roundabout beside the police headquarters in Monkmoor (north-east Shrewsbury), take the road which leads north-east to pass Monkmoor Farm. The road becomes unmetalled and somewhat pot-holed, but persist until reaching a car parking space just short of the river. The pool — not shown on maps — is easily seen from here, and riverside paths run in both directions.

## Calendar

*Resident:* Mallard, Sparrowhawk, Kestrel, Moorhen, Coot, Kingfisher, Green Woodpecker, Great Spotted Woodpecker, Grey Wagtail, Reed Bunting.

*April–June:* Passage waders and terns. Summer migrants from mid-April include Cuckoo, Yellow Wagtail, Sedge Warbler. Swift from early May.

*July–September:* Passage waders and terns. Finch flocks forming. Early single Cormorants.

*October–March:* Little and Great Crested Grebe move to river. Cormorant. Winter wildfowl include more on river in freezing spells. Fieldfare, Redwing, finch flocks, including Goldfinch. Sand Martins arrive late March.

# HAUGHMOND HILL

## Habitat

Haughmond Hill is a low plateau rising 250 ft (77 m) above the surrounding plain, with rock outcrops and steep wooded slopes on the western and southern facing boundaries. It lies 3 miles (4.8 km) east of Shrewsbury. Although large sections of the hill have been afforested with regimented conifers, there are enough broad-leaved areas remaining to provide a variety of habitats, although there is now much less open space. A forest fire in the 1970s destroyed many acres of coniferous woodland, creating for a while a habitat suitable for Grasshopper Warblers, Whinchats and Stonechats, but not, as was hoped, attracting Nightjars back into an area where they were regular breeding birds not too long ago. However, the improvement was short-lived, and the area was quickly replanted. A vagrant Red-footed Falcon was tempted to spend a fortnight in this replanted area in June 1982, hunting regularly above the infant conifers. The prolific growth rate of these trees has resulted in a rather dull landscape today.

The most interesting woodlands lie on the western side of the hill, from Haughmond Abbey in the north to Downton in the south, where broad-leaved trees still dominate, including many oaks. These occur at low level and along the slopes. There are areas of birch wood and scrub at high level along this side, and a few Scots pines. A quarry in the north-west of the hill has been responsible for the destruction of some habitat, but a good area of mature oaks remains to the east of the workings.

## Species

A walk through the low-level woods along the western fringe of Haughmond Hill will reveal resident woodland birds at all times of the year, and all three woodpeckers may be seen, along with Nuthatches and Treecreepers. Long-tailed and other tits form mixed roving bands in autumn, and in summer the woods are busy with warblers, especially Blackcaps and Willow Warblers. All the broad-leaved areas are good for Sparrowhawks, Tawny Owls, Redstarts, Spotted Flycatchers, occasional Pied Flycatchers, Jays and other common woodland passerines, whilst also providing many sites for hole-nesting species such as Stock Doves, Jackdaws, Starlings and tits. Kestrels are regulars on the hill, hunting the remaining open areas. A few pockets of rough grass can still be found where Grasshopper Warblers are heard and in scrubby growth, or among newly planted conifers, Whinchats, Whitethroats, and even occasional Stonechats are seen. The hill is good generally for small passerines such as Wrens, Dunnocks, Robins, resident thrushes and finches, and Yellowhammers, whilst winter months will produce Fieldfares, Redwings, Siskins and Redpolls. The coniferous plantations contain Goldcrests and Coal Tits. Tree Pipits breed in some numbers, and may play host, along with Dunnocks and other small birds, to Cuckoos. A leading feature of the ornithological year is provided by roding Woodcocks, which invariably give excellent views in spring and summer. Hobbies hunt the hill in some years, especially during post-breeding dispersal. Swifts, sometimes in large numbers, hunt above the hill in summer months.

## Timing

Haughmond Hill is a local beauty spot and quite well used by walkers. It is not really a tourist attraction, hence disturbance never reaches chronic proportions. Summer weekends are the periods of maximum disturbance. There is only a minimum of birdlife in winter, but visits in spring and early summer can be very rewarding, especially after the arrival of most migrants by late April, but before the foliage of high summer makes the viewing of woodland birds difficult. Early morning is the best time of day to visit during this period, although evenings have their own special magic at the approach of dusk, with Woodcocks roding, Tawny Owls hooting, bats on the wing and a good chance of seeing fallow deer.

## Access

Haughmond Hill can best be approached from Shrewsbury by taking the B5062 which leaves the A49(T) at a large roundabout 1¼ miles (2 km) north of the town centre. An alternative route is to take the unclassified road which leaves the A5(T) 4 miles (6.4 km) south-east of the town, alongside Attingham Park. Limited roadside parking space near Haughmond Abbey, on the B5062 north of the hill, enables walkers to explore the western woods at low-level before heading for higher ground. Another access point is by a track from the minor road which leaves the B5062 at a T-junction 300 yds (277 m) to the east of the abbey entrance. There are parking spots along this road. Roadside parking is also available on the east, just south of The Criftin, where a forest road leads to the coniferous plantations, and on the minor road to the south-west, from which two public rights of way cross fields to reach the hill. An attractive approach is along a strip of woodland from the north of Uffington, but parking in the village may prove difficult.

There is no vehicular access to the hill, nor is there unlimited open access on foot. However, there are good tracks and paths in addition to some rights of way.

## Calendar

*Resident:* Sparrowhawk, Kestrel, Woodcock, Tawny Owl, the three woodpeckers, Nuthatch, Treecreeper. Resident thrushes, tits, crows, finches.

*April–June:* Roding Woodcock. Summer visitors from mid-April include Cuckoo, Tree Pipit, Redstart, Grasshopper Warbler, scrub and leaf warblers and occasional Pied Flycatcher. Turtle Dove and Spotted Flycatcher from early May.

*July–September:* Occasional Hobby. Flocks of mixed juvenile tits. Finch flocks form. Jays carry acorns.

*October–March:* Stonechat, Fieldfare, Redwing, finch flocks including Siskin and Redpoll.

## EARL'S HILL AND PONTESFORD HILL

### Habitat

These two hills, almost one feature with separate names rather than two distinct hills, lie south of the A488 7½ miles (12 km) to the south-west of Shrewsbury. The wild grandeur of Earl's Hill, a pre-Cambrian pile reaching 1,040 ft (320 m), forms the southern half of the area, and the heavily afforested Pontesford Hill makes up the remainder. Earl's Hill is a superb spot, with a rich variety of natural features, and is a nature reserve of the Shropshire Trust for Nature Conservation (STNC). From the summit the views are magnificent, and it is easy to see why Iron Age Man chose the site as a vantage point and to provide protection. Traces of the hill fort are still clearly discernible today, two thousand years on.

The summit of Earl's Hill is surrounded by open, rough grassland, which also covers the north-west slopes descending to the Craft Valley, there to meet the regimented conifers of Pontesford Hill. The drier grassland on the rockier southern slopes has been invaded by broom, and at lower levels by a thicket of scrub and bracken. Mainly ash, elder and hawthorn, with bramble, this is perfect warbler habitat. East of the summit a dramatic rock face towers above scree slopes and a deep wooded valley, through which the Habberley Brook flows, tumbling between boulders and cascading over shelves of bedrock to provide ideal conditions for Dippers and Grey Wagtails. Alders grow along the stream, with the hillside wood containing mainly ash, wych elm and oak. Trees also grow in the gullies which cut into the crags, reaching almost to the summit in places. The scree slopes are enlivened with rock stonecrop, and common lizards bask here on sunny days.

The great strength of Earl's Hill as a nature reserve lies in the rich mosaic of habitats crammed into a relatively small area. It is well known for the impressive number of butterfly species to be seen there.

Pontesford Hill is much less exciting, due to the blanket of conifers which cover most of the surface. Even so the woodland edges, a few mixed areas, and some small but more open patches at the northern end, provide a good selection of woodland passerines.

The hills are surrounded by agricultural land on three sides, with extensive woodland beyond the Habberley Brook on the east. There is an excellent Visitor Centre in the painstakingly restored barn, sited above the valley and approached by crossing interesting hill meadows to the north-east of the reserve.

### Species

Although a visit during any season will be rewarded with some worthwhile sightings, it is in spring and summer, when the woods are vibrant with bird song, butterflies are on the wing, and carpets of flowers and blossoms splash colour across the scene, that the area shows its true splendour.

In the broad-leaved woods, Wood Warblers trill, six species of tits perform the hectic chores of their breeding cycle, and all three woodpeckers may be heard and seen. Treecreepers and Nuthatches, so often overlooked, search the boles for insect food, whilst in more open areas Tree Pipits perform their parachute flight. Sit on a high vantage point and look down onto the woods for a while and the brilliant flash of a Jay may be glimpsed momentarily, or the quick dash of a hunting Sparrowhawk. The flickering white patches of Pied Flycatchers attract the eye, and Redstarts are betrayed by their fiery tails, although the white forehead of the cock is always surprisingly conspicuous. Turtle Doves croon, Chiffchaffs sing monotonously from tree-top song posts, and Cuckoo calls echo across the valley. Buzzards soar over these woods, and a cacophony of angry alarm cries may betray the presence of a roosting Tawny Owl, discovered perhaps by a Blackbird whose noisy objections lead the chorus of dissent.

The coniferous woods are less productive, but Goldcrests and Coal Tits feed here. The former also breed, but the hole-nesting tits must move out to raise their broods. Siskins visit the woods in winter.

In scrubby places, Whitethroats, Garden Warblers and Blackcaps are found, together with the ubiquitous Willow Warbler. Linnets and Yellowhammers find this habitat to their liking, whilst in winter Redpolls feed in the birches. In rough grassy patches on the edges of scrub, and towards the southern end of the valley, Grasshopper Warblers are heard though seldom seen, especially at close of day, when Woodcocks are roding overhead.

The stream is closely hemmed in by trees, making it difficult to watch along anything but a short stretch. It is frequented by Dippers and Grey Wagtails, although Kingfishers are rarely seen. Herons are surprised here occasionally, and Mallards, but the flicking white undertail coverts of Moorhens are usually glimpsed only briefly as a bird swims jerkily for cover, or runs crouching under the stream banking.

Kestrels hunt the open hillside grassland, and Meadow Pipits occur in some numbers, with a few Skylarks. In the meadows south of the Visitor Centre, ground-feeding Green Woodpeckers seek their favoured diet of ants and small charms of Goldfinches visit thistles.

Rock climbers scale the crags during weekends and summer evenings, but are tolerated by noisy Jackdaws which nest in the cliff crevices. Redstarts also find nest-sites in the tree-clad gullies. In undisturbed periods Ravens call from the rocks and Swifts scream noisily around the brooding faces in an endless search for insects.

## Timing

Spring, with resident breeding birds established on territory, summer migrants newly arrived, and broad-leaved trees not yet in full foliage, is the best time of the year. Early morning is the best time of the day for maximum bird activity, but evenings can also be good, with reeling Grasshopper Warblers, roding Woodcocks, a late hunting raptor or two, and bats on the wing. With care, badgers can be seen and gloworms have been noted in the dusk of summer evenings. For soaring Buzzards mid-day is the best time to visit.

## Access

Take the A488 Bishop's Castle road from Shrewsbury, and park at the garage on the left, 6 miles (9.6 km) after crossing the A5(T) and ½ mile (0.8 km) before reaching Pontesbury. Visitors are especially asked not to park in the narrow lane at the north of Pontesford Hill, as this could impede the access of emergency vehicles in the event of a fire or a climbing accident.

Follow the lane from the garage (ignoring the turning to the left after 550 yds (500 m) to reach a wide footpath leaving the road on the left) which is the most direct route to the scree slopes, Habberley Brook and hillside woodland. A short diversion, after passing through a field gate

195

into a meadow, leads to the Visitor Centre (signposted). The centre has restricted opening hours, but is manned during most summer weekends, and has a permanent exhibition giving information about the reserve, and a nature trail which has been set out to cover the most interesting features of the site. The area is well served with footpaths.

## Calendar

*Resident:* Sparrowhawk, Buzzard, Kestrel, Tawny Owl, all three woodpeckers, Nuthatch, Treecreeper, Dipper, Goldcrest, six tits, other resident woodland passerines, resident finches, Yellowhammer.

*April–June:* Summer migrants arrive, mostly from mid-April, including Turtle Dove, Cuckoo, Swift, Tree Pipit, Redstart, Grasshopper Warbler, scrub warblers, leaf warblers, Spotted and Pied Flycatchers. Woodcock roding. Soaring raptors. Many breeding species.

*July–September:* Swifts and hirundines feed around crags in large numbers. Finch flocks forming and large bands of juvenile Blue and Great Tits, accompanied by smaller numbers of other tits, Goldcrests and Willow Warblers, feeding in woods. Jays carrying acorns.

*October–March:* Fieldfares and Redwings from mid-October feeding on berries. Siskins in conifers and joining Redpolls in birch and alder. Tawny Owls disputing territories at beginning of period. Goldcrests and Coal Tits in conifer woods. Resident species take up territories at end of period. Chiffchaff arrives late March.

# THE NORTH SHROPSHIRE MERES

Map 51
OS map 126

## Habitat

The meres spread south-east from Ellesmere, forming a miniature Lake District in the low-lying north Shropshire countryside. A total of eight meres lie within a compact area of 3 miles (5 km) by 2½ miles (4 km), the most accessible being The Mere, Blake Mere, Cole Mere, Newton Mere and White Mere. Hammer Mere, 4½ miles (7.2 km) north-east of Ellesmere, is also worth visiting, although it lies beyond the Shropshire boundary in Clwyd.

The meres were formed by glaciation. Terminal moraine deposited during the last Ice Age left an uneven surface, hollows in which were filled by melting ice. The retreating ice also left behind isolated blocks which melted into steep sided kettleholes. The result is the delightful, gently undulating, lake-dotted area we see today. Not being fed by rivers or streams, the meres have a water level corresponding with the surrounding water table, resulting in little variation in depth over the year. This makes them particularly good for waterfowl, but less attractive to waders due to the absence of exposed mud which falling water levels would provide. Most of the meres contain good populations of fish, making them attractive to grebes, Cormorants, Grey Herons, sawbills and the occasional wintering diver.

Two man-made features add to the ornithological interest of this area. The Shropshire Union Canal runs alongside Blake Mere and Cole Mere, linking the two with a waterway passing through interesting habitats of woodlands, open fields and damp patches of rough herbage. At dusk on spring and summer evenings, bats abound, the flittering pipistrelle, and Daubenton's bat skimming rapidly just above the water surface, being easiest to identify. Wood Lane gravel pit, between White Mere and Cole Mere, provides suitable habitat for nesting Sand Martins and is a good staging post for passage waders.

Between the meres, the land is given over to agriculture, mainly cereals and pasture, but with generous pockets of mixed woodlands, ideal for woodpeckers and small passerines.

## Species

Many of the common species are found on most of the meres and it is usually quite easy to locate Little and Great Crested Grebes, Cormorants in winter, Mute Swans, feral Greylag Geese, Canada Geese, Mallard, Tufted Duck, Ruddy Duck, wintering Wigeon, Teal, Shoveler, Pochard, Goldeneye, Goosander, Moorhens, Coots and the common gulls. Less usual species which occur annually and may appear on any water are Bewick's Swans, White-fronted and Barnacle Geese, Shelduck, Pintail, Smew, Common, Arctic and Black Terns. Other birds arrive less frequently, perhaps storm driven by westerly winds, or as part of a cold weather movement due to harsh conditions further north. These include Red-throated Divers, Black-necked Grebes, Whooper Swans, Bean and Pink-footed Geese, Gadwall, Scaup, Long-tailed Duck, Common Scoter and Red-breasted Merganser. Scarcer still are Black-throated and Great

197

Northern Divers, Red-necked Grebes, Brent Geese and Feruginous Duck. Some wildfowl are difficult to categorise, occurring in Britain as wild birds, but being more often met with as escapees, and the meres have hosted Snow Geese, Ruddy Shelduck and Mandarin in recent years. Should you come across such species, do not worry about their status, simply enjoy them, for they are all handsome birds. Water Rails are heard in winter, and most meres have suitable habitat for them, but the southern tip of The Mere is as good a place as any to look.

*Young Great Crested Grebes can be seen on the Shropshire meres in summer*

Some species have definite preferences for particular meres. The Mere has a small heronry, with up to twelve nests, on the island. This large expanse of water holds a large gull roost, building up from late summer to maximum numbers in mid-winter. Black-headed Gulls dominate, sometimes making up over three-quarters of the total, which reaches 15,000 birds. Lesser Black-backed, Herring and Common Gulls make up the numbers. Great Black-backed Gulls rarely reach a dozen, and Kittiwakes are unusual visitors. Scarcer gulls are now being recorded annually, with Little Gulls arriving in late summer, followed by Mediterranean Gulls from late autumn, but Iceland and Glaucous Gulls are not usually present until the turn of the year. The Mere usually takes the very few terns on spring passage, and many of those on the return autumn journey, when birds are more numerous and longer staying. Common Terns arrive first, from mid-August, followed by Arctic Terns in early September. Black Terns, always the most recorded, have a protracted passage from late August to late October. Single birds are the norm, with up to five on occasions. The alders on the western corner of The Mere often contain feeding Siskins in winter, and the wood on the north-west shore is good for Lesser Spotted Woodpeckers.

Cole Mere also holds a large gull roost with up to 10,000 present in January, and is also an excellent mere for passage terns. This is the best water to see small numbers of White-fronted Geese in winter; and a half-dozen Barnacle Geese bored with life in some wildfowl collection, have made it their home since 1982, although they do visit other meres. The marshy area at the open south-east end of the mere holds good numbers of winter Snipe, with one or two Jack Snipe in some years. The same area is graced by Yellow Wagtails in spring and summer.

Blake Mere holds tree-roosting Cormorants, and both Newton Mere and White Mere are favoured sites of storm-driven divers. Crose Mere, viewable only from a distance, is a gathering ground for post-breeding Ruddy Ducks before they move across to the West Midland reservoirs for the winter, and numbers have topped 100 in recent autumns. Hanmer Mere serves the same purpose for these now well-established stifftails, and often holds up to 60 wintering Goosander.

The Mere, Blake Mere and Cole Mere have mixed woodland adjacent to their shores in which can be found Sparrowhawks, Turtle Doves, Tawny Owls, the three woodpeckers, Goldcrests, most of the common woodland passerines and Jays.

The Wood Lane sand and gravel pit can be viewed from the road on its northern boundary. It has a Sand Martin colony, attracts a few wildfowl, including Shelduck, and gulls — the only Sabine's Gull recorded in Shropshire was seen here in July 1983 — but is best known for passage waders. The spring movement is always much weaker than the return in autumn, and often limited to half-a-dozen species. Dunlins, normally the first arrivals, pass through in small flocks, but a flock of 45 was counted in late April 1985. Oystercatchers, in groups of up to four, are always well represented and pairs have summered in the area some years. Little Ringed and Ringed Plovers, together with Redshanks and Common Sandpipers usually appear as singles. Curlews gather here from late spring to mid-autumn in noisy bathing and preening parties, or later as post-breeding flocks which may top 120 birds. Whimbrels and Black-tailed Godwits also call here in spring. First of the return passage to arrive are Ringed Plovers, Dunlins and Redshanks, seen from mid-July, and they are joined in early August by Oystercatchers, Ruffs, Greenshanks and Common Sandpipers. Green and Wood Sandpipers are also seen annually, with Whimbrels, Knots and Little Stints passing through in some years.

The stretch of the Shropshire Union Canal between Blake Mere and Colemere can be very rewarding. It gives the best access to Blake Mere, where Little and Great Crested Grebes, Cormorants and a few wildfowl regularly occur. To the east of this mere the woodland is waterlogged, containing many old and decaying trees which often play host to Lesser Spotted Woodpeckers and Redstarts. Moorhens nest in emergent vegetation alongside the towpath, and Kingfishers are not infrequent. It is a strange sight to see them disappearing into, or emerging from, the long tunnel to the west of Blake Mere. Winter thrushes in large numbers feed in the fields and hedgerows north of the canal, and the woods on the southern bank produce much bird activity. On spring evenings the melancholy hooting of Tawny Owls echoes from the woodland depth, with the strident calls of Little Owls from the edges.

## Timing

The meres are very much a tourist area, and birdwatchers may find too many people for their liking during summer weekends and Bank Holidays. The canal is used by pleasure craft, especially in these periods. However, most visitors stay in the area of The Mere, and there is a beneficial side effect in the way the birds become accustomed to human presence. The Herons continue to nest on the island opposite the main car park, and Canada Geese, Mallards and Coots walk freely among admiring tourists. Even in busy periods an early morning or late evening visit will give quiet conditions. Gull roosts start to form from mid-afternoon, with the majority of birds arriving before dusk. During the short winter days there is constant bird activity from dawn to dusk, and this is the most rewarding time to visit the area. Weather should be taken into account, and a visit following westerly gales may be rewarded with a rarity or two. Long periods of freezing weather leave little or no unfrozen water, and wildfowl move out of the area.

## Access

There is no access to Crose Mere or the nearby tiny Sweat Mere, which lie 1½ miles (2.5 km) to the south of the main group, although the former can be seen from the A528 to the west. However, the mere is fringed by tall trees and viewing from here is unsatisfactory. A public right of way passes between the two meres, but gives no access to the shores.

Three other meres can only be viewed from the road: White Mere, alongside the A528, can best be seen from the minor road to the north; Newton Mere is viewable from the minor road passing the southern tip; and Kettle Mere can be seen from the A495. The former two should not be missed, especially after westerly gales. Another roadside site is Wood Lane gravel pit, which lies on the southern side of the minor road from the A528 to Colemere.

The Mere has official car parking on the south-west shore alongside the A528, and limited roadside parking on the north-west shore. There is open access from the southern tip for about three-quarters of the perimeter, until a meadow halfway along the north-east shore is reached, at which point it becomes private property. Limited space for parking at the junction of the A528 and A495 gives access to the canal towpath, from which Blake Mere can be seen, and from which Cole Mere may be reached by birdwatchers wanting an interesting and picturesque walk. Cole Mere is owned by Shropshire County Council (SCC), which has provided a car park on the Colemere to Lyneal road beside the southern end of the mere. The Council has also provided a super footpath which circumnavigates the mere.

Hanmer Mere is reached by driving along the A495 Ellesmere to Whitchurch road and taking a minor road north-west after 5 miles (8 km). There is a public right of way along the eastern shore.

## Calendar

*Resident:* Little Grebe, Great Crested Grebe, Mute Swan, Canada Goose, Mallard, Tufted Duck, Moorhen, Coot, Kingfisher, Sparrowhawk, Kestrel, Little Owl, Tawny Owl, the three woodpeckers and common resident passerines.

*April–June:* Young Herons stand on nests from mid-May. Passage waders at Wood Lane include Oystercatcher, Dunlin, Little Ringed Plover, Ringed Plover, Redshank and Common Sandpiper. Shelduck, Common and Arctic Terns. Most summer migrants from mid-April, but Swift, Turtle Dove and Spotted Flycatcher from early May.

*July–September:* Passage waders at Wood Lane including Dunlin, Ruff, Redshank, Greenshank, Green, Wood and Common Sandpipers.

Common, Arctic and Black Terns on meres. Possibility of Little Gulls. Post-breeding flocks of Curlew at Wood Lane. Passage Yellow Wagtail, Cole Mere.

*October–March:* Divers (scarce) usually from January after storms. Main Cormorant arrival from mid-October. Herons breeding from mid-February. Ruddy Duck gather on Crose Mere and Hanmer Mere in October. Main arrivals of winter wildfowl from late-October or early-November, including Wigeon, Teal, Mallard, Shoveler, Pochard, Tufted Duck, Goldeneye and Goosander (especially Hanmer Mere) in good numbers, but also Bewick's Swan and scarcer Whooper Swan, White-fronted Goose, Pintail and Smew. In some years Bean Goose, Pink-footed Goose, Brent Goose, Gadwall, Scaup, Long-tailed Duck, Common Scoter and Red-breasted Merganser. Goldeneye gather on The Mere, late March. Water Rail (scarce). Gull roost, maximum numbers December–January including Mediterranean (from October), Iceland and Glaucous Gulls. Snipe and occasionally Jack Snipe, Colemere. Winter thrushes and finches include Siskin and Redpoll. Chiffchaff and Sand Martin arrive late March.

# LIST OF ADDITIONAL SITES

In order to deal adequately with each site, we have had to be selective in those we have described. Inevitably many sites have had to be omitted. In the main, areas where access is difficult or the habitat is liable to change have not been included. Nor have isolated sites which on their own are unlikely to provide a good day's birdwatching. The following additional sites, however, may be worth visiting.

| Site Grid Ref | Habitat | Species | Timing | Access |
|---|---|---|---|---|
| Allscott Sugar Factory SJ 605 127 | Settling pools, river and marsh | Wildfowl, waders, terns and aquatic passerines | All, but especially spring and autumn | Strictly permit only to SOS members |
| Berkswell SP 240 791 | Parkland and lake | Common wildfowl, Yellow Wagtail, Grasshopper Warbler, Heronry nearby | Spring, summer | Public footpath from Berkswell Church |
| Bircher Common SO 457 665 and Leinthall Common SO 450 674 | Grass and bracken hillsides, scrub and mixed woodland | Raptors, pipits, common scrub and woodland birds | Spring, summer | Bircher Common NT: open access Leinthall Common public footpaths |
| Bishops Wood SJ 750 310 | Coniferous woodland with broad-leaved fringe | Common woodland birds include Tree Pipit. Winter thrush roost | Spring | Public footpaths through wood |
| Branston SK 212 205 | Gravel Pit | Wildfowl, waders | Spring, autumn winter | View from public rights-of-way |
| Brown Moss SJ 663 395 | Deciduous woodland, pools, marsh and heathland | Grey Heron, wildfowl, Water Rail, Reed Bunting and common scrub and woodland birds | All (avoid freezes) | View from roads and footpaths |
| Chaddesley Wood SO 914 736 | Mixed woodland | Common woodland birds | Spring | NNR: follow waymarked paths |
| Chesterton SP 357 582 | Pools and farmland | Common wildfowl, perhaps rare water bird | Spring, winter (avoid freezes) | Public footpath from Chesterton Church |
| Clent Hills SO 940 796 | Hills with grass, bracken, scrub and mixed woodland | Common passerines, passage upland birds, Dipper, Grey Wagtail | Spring, summer, autumn | Several car parks and general access |
| Coombe Abbey SP 404 798 | Parkland, lake and mixed woodland | Heronry, wildfowl (esp. Shoveler), Woodcock, owls, Kingfisher, common woodland birds | All | Country Park: car park and general access (charge made) |
| Copmere SJ 805 295 | Glacial lake, reedbed, woodland | Wildfowl (more in winter), Water Rail, Kingfisher, Grey Wagtail, Reed Warbler, common woodland birds | All | View from adjoining lanes |
| Crewe Green SJ 330 158 | Extensive river floods | Wildfowl, including wild swans | Winter | View from road |
| Croft Castle SO 463 655 | Parkland and mixed woodland | Raptors, Pied Flycatcher, common woodland birds | Spring, summer | NT: nature walks and footpaths |

# LIST OF ADDITIONAL SITES

| Site Grid Ref | Habitat | Species | Timing | Access |
|---|---|---|---|---|
| Crinshill including Corbet Wood SJ 519 237 | Sandstone cliffs, heath and mixed woodland | Sparrowhawk, Kestrel, Woodcock, common woodland birds | All | Public footpaths STNC nature trail in Corbet Wood |
| Devils Spittleful SO 807 747 and Hartlebury Common SO 820 705 | Lowland heaths with some woodland | Woodcock, Green Woodpecker, Tree Pipit, Stonechat, Whitethroat, Hawfinch, commoner passerines. Thrushes and Siskin in winter | Spring, winter | Devils Spittleful: WNCT public path crosses area. Hartlebury: LNR general access on either side of A4025 |
| Downs Banks SJ 902 366 | Bracken hillsides, deciduous woodland, stream | Common passerines, Grey Wagtail, roosting thrushes in winter | Spring, winter | NT: general access from adjacent lane |
| Earlswood Lakes SP 114 742 and Clowes Wood SP 098 738 | Lakes, deciduous woodland | Common wildfowl (mostly in winter), woodland birds including Wood Warbler | Spring, summer, winter | Public footpaths around lakes and through woods. Clowes Wood: WARNACT reserve members only |
| Edge Hill SP 380 480 and Burton Dassett Hills SP 395 520 | Ironstone hills. Edge Hill: deciduous woodland. Burton Dassett: grass and scrub | Common passerines | Spring | General access from adjacent roads |
| Elford SK 184 085 | Gravel pits | Wildfowl, waders, breeding Black-headed Gull | Spring, autumn, winter (avoid freezes) | View from adjacent road |
| Enville SO 836 865 and Kinver SO 834 830 | Heaths with coniferous woodland at Enville and oak-birch woods at Kinver | Common heath and woodland species, perhaps also Long-eared Owl, Nightjar and Crossbill | Spring to early autumn | General access on either side of road at Enville and rides through FC plantations. General access to hills at Kinver (NT and Country Park) |
| Gailey SJ 937 103 | Reservoirs | Heronry, winter wildfowl, Cormorant roost in winter | Spring, winter | Can be partly viewed from adjacent roads, otherwise permit from WMBC |
| Goosehill Wood SO 935 608 | Scrub woodland | Common scrub and woodland birds, Nightingale | Spring | WNCT reserve: members only |
| Greenway Bank and Knypersley Pool SJ 890 551 | Deciduous woodland and lakes | Wildfowl (most winter), Grey Wagtail, Dipper, common woodland birds | Spring, autumn, winter | Country Park: car parks and general access (charge made) |
| Grimley SO 840 607 and Holt SO 826 622 | Gravel pits and marsh | Wildfowl (most winter), passage waders, passerines | Spring, autumn, winter (avoid freezes) | View from roads and public footpaths |
| Hanchurch Hills SJ 840 397 | Coniferous woodland | Common woodland birds | Spring | FC: public footpaths |

# LIST OF ADDITIONAL SITES

| Site Grid Ref | Habitat | Species | Timing | Access |
|---|---|---|---|---|
| Highgate Common SO 843 900 | Heath, birch and conifer woodland | Common heath and woodland birds | Spring | Country Park: car parks and general access |
| Hunthouse Wood SO 705 702 | Deciduous woodland in steep valley | Raptors, common woodland birds | Spring | WNCT reserve: members only |
| Hurcott Pool SO 851 779 | Millpool | Common wildfowl, Kingfisher | Winter (avoid freezes) | View from adjacent road |
| Kings Bromley SK 111 167 | Gravel pit and river | Cormorant, wildfowl (including Goosander) | Winter (avoid freezes) | View from road |
| Larford SO 816 692 | Gravel pit | Wildfowl (most winter), passage waders | All (avoid freezes) | View from road |
| Llanymynech SJ 266 218 and Llynclys SJ 273 237 | Limestone quarry, scrub, heath, grassland and woodland | Raptors (incl occasional Peregrine) Tree Pipit, Redstart, Pied Flycatcher, Raven, common woodland birds | Spring, summer | STNC reserves: permits not required |
| Merrington Green SJ 465 209 | Scrub, woodland, pools and grassland | Sparrowhawk, Kestrel, common passerines | Spring, summer | STNC: nature trail |
| Monkwood SO 804 607 | Mixed deciduous woodland and scrub | Common woodland birds plus Nightingale, Grasshopper Warbler: winter roost | Spring, autumn, winter | View from road through wood |
| Nescliffe SJ 386 195 | Sandstone outcrop, scrub, heath and mixed woodland | Small raptors, passage Wheatear common scrub and woodland passerines | Spring, summer | Public footpaths |
| Newbold Comyn SP 335 655 | Flood meadows, fields and hedgerows | Snipe, common passerines: a few wildfowl when flooded | Winter, spring | Car parks and general access |
| Old Hills Common SO 830 488 | Common grazing with gorse and scrub | Common scrub birds: perhaps Nightingale | Spring | General access either side of road |
| Ravenshill Wood SO 739 539 | Mixed woodland | Woodcock, warblers, other woodland birds | Spring | WNCT reserve: members only |
| Seeswood Pool SP 329 904 | Subsidence pool | Wildfowl, passage waders, gulls and terns | Spring, autumn, winter (avoid freezes) | View from adjacent road |
| Shobdon SO 402 621 | Pools, mixed woodland, disused airfield | Few wildfowl (most winter), Golden Plover (winter), common woodland birds | All (avoid freezes) | View from roads and public footpaths |
| Shrawley Wood SO 810 655 | Mixed woodland, pools and stream | Kingfisher, Grey Wagtail, Dipper, common woodland birds | Spring | Large part FC: Keep to public footpaths |

204

| Site Grid Ref | Habitat | Species | Timing | Access |
|---|---|---|---|---|
| Soudley Wood<br>SO 330 806 | Mixed hillside woodland, conifer plantations | Raptors, Woodcock, Tawny Owl, woodpeckers and many passerines including Pied Flycatcher, Redstart and Crossbill | Spring, summer | Public footpaths |
| Stratford-on-Avon Canal<br>SP 190 676 | Canal, stream and wet meadows | Snipe, Kingfisher, Grey Wagtail, Siskin | Winter | NT: along towpath |
| Tiddesley Wood<br>SO 933 449 | Mixed woodland and stream | Nightingale, common scrub and woodland birds | Spring, winter | WNCT reserve: members only |
| Venus Pool<br>SJ 549 061 | Shallow pool, emergent vegetation | Wildfowl, passage waders and terns, Black-headed Gull colony | Spring, autumn, winter | SOS reserve: members only |
| Whichford Wood<br>SP 298 341 | Mixed woodland | Common woodland birds | Spring | Footpaths from nearest road |
| Whittington Sewage Farm<br>SO 872 839 | Sewage works and fields | Snipe, Curlew, passerines: roost site | Winter | View from road |
| Wormleighton Reservoir<br>SP 449 518 | Canal-feeder reservoir | A few wildfowl, waders | Autumn, winter | Along towpath |

# USEFUL ADDRESSES AND PUBLICATIONS

The following addresses may be useful in planning your visit. We also include a list of publications which birdwatchers might find particularly useful, but we have not attempted to compile a comprehensive bibliography.

## Addresses

British Trust for Ornithology,
Beech Grove,
Tring,
Hertfordshire.

Forestry Commission,
North-west Region,
Dee Hills Park,
Chester CH3 5AT.

Herefordshire Ornithological Club,
Secretary, Mrs J.M. Bromley,
The Garth,
Kington,
Herefordshire HR5 3BA.

Herefordshire and Radnorshire
    Nature Trust,
25 Castle Street,
Hereford HR1 2NW.

National Trust,
Severn Regional Office,
34 Church Street,
Tewkesbury,
Gloucestershire GL20 5SN.

Nature Conservancy Council,
West Midlands Regional Office,
Attingham Park,
Shrewsbury SY4 4TW.

Royal Society for the Protection
    of Birds,
Midlands Regional Office,
44 Friar Street,
Droitwich,
Worcestershire WR9 8ED.

Shropshire Ornithological Society,
Secretary, H.J. Blofield,
'Arnsheen',
Betley Lane,
Bayston Hill,
Shrewsbury SY3 OAS.

Shropshire Trust for Nature
    Conservation,
Conservation Officer, J.J. Tucker,
St George's School,
New Street,
Shrewsbury SY3 8JP.

Staffordshire Nature Conservation
    Trust,
Administrative Officer, H. Haseley,
3a Newport Road,
Stafford ST16 2HH.

Urban Wildlife Group,
11 Albert Street,
Birmingham B4 7UA.

Warwickshire Nature Conservation
    Trust,
Montague Road,
Warwick CV34 5LW.

West Midland Bird Club,
PO Box 1,
Studley,
Warwickshire B80 7JG.

Worcestershire Nature Conservation
    Trust,
Hanbury Road,
Droitwich,
Worcestershire WR9 7DU.

# Publications

Herefordshire Ornithological Club Annual Report, which covers Herefordshire and Radnorshire. Published by Herefordshire Ornithological Club.

Shropshire Bird Report. Published annually by the Shropshire Ornithological Society.

West Midland Bird Club Annual Report, which covers Warwickshire, Worcestershire, Staffordshire and the former West Midlands County. Published by the West Midland Bird Club.

Harrison, G.R., A.R. Dean, A.J. Richards and D. Smallshire (1982) *The Birds of the West Midlands*, WMBC Studley.

Ratcliffe, Dr. D.A. (ed.) (1977) *A Nature Conservation Review*, Cambridge University Press, Cambridge.

Rutter, E.M., F.C. Gribble and T.W. Pemberton (1964) *A Handlist of the Birds of Shropshire*, SOS Shrewsbury.

Voous, Dr K.H. (1977) *List of Recent Holarctic Bird Species*, London.

Walker, C.W. and A.J. Smith (1975) *Herefordshire Birds*, Woolhope Naturalists' Field Club, Hereford.

# LIST OF BIRDS MENTIONED IN TEXT

The sequence follows Professor K.H. Voous's *List of Recent Holarctic Bird Species* (1977).

| English Name | Scientific Name |
|---|---|
| Red-throated Diver | *Gavia stellata* |
| Black-throated Diver | *Gavia arctica* |
| Great Northern Diver | *Gavia immer* |
| Little Grebe | *Tachybaptus ruficollis* |
| Great Crested Grebe | *Podiceps cristatus* |
| Red-necked Grebe | *Podiceps grisegena* |
| Slavonian Grebe | *Podiceps auritus* |
| Black-necked Grebe | *Podiceps nigricollis* |
| Fulmar | *Fulmarus glacialis* |
| Cory's Shearwater | *Calonectris diomedea* |
| Manx Shearwater | *Puffinus puffinus* |
| Storm Petrel | *Hydrobates pelagicus* |
| Leach's Petrel | *Oceanodroma leucorhoa* |
| Gannet | *Sula bassana* |
| Cormorant | *Phalacrocorax carbo* |
| Shag | *Phalacrocorax aristotelis* |
| Bittern | *Botaurus stellaris* |
| Little Bittern | *Ixobrychus minutus* |
| Night Heron | *Nycticorax nycticorax* |
| Little Egret | *Egretta garzetta* |
| Grey Heron | *Ardea cinerea* |
| Purple Heron | *Ardea purpurea* |
| Black Stork | *Ciconia nigra* |
| Spoonbill | *Platalea leucorodia* |
| Mute Swan | *Cygnus olor* |
| Bewick's Swan | *Cygnus columbianus* |
| Whooper Swan | *Cygnus cygnus* |
| Bean Goose | *Anser fabalis* |
| Pink-footed Goose | *Anser brachyrhynchus* |
| White-fronted Goose | *Anser albifrons* |
| Greylag Goose | *Anser anser* |
| Snow Goose | *Anser caerulescens* |
| Canada Goose | *Branta canadensis* |
| Barnacle Goose | *Branta leucopsis* |
| Brent Goose | *Branta bernicla* |
| Ruddy Shelduck | *Tadorna ferruginea* |
| Shelduck | *Tadorna tadorna* |
| Wood Duck | *Aix sponsa* |
| Mandarin | *Aix galericulata* |
| Wigeon | *Anas penelope* |
| Gadwall | *Anas strepera* |
| Teal | *Anas crecca* |
| Mallard | *Anas platyrhynchos* |
| Pintail | *Anas acuta* |
| Garganey | *Anas querquedula* |
| Blue-winged Teal | *Anas discors* |
| Shoveler | *Anas clypeata* |
| Red-crested Pochard | *Netta rufina* |
| Pochard | *Aythya ferina* |
| Ring-necked Duck | *Aythya collaris* |

| English Name | Scientific Name |
|---|---|
| Ferruginous Duck | *Aythya nyroca* |
| Tufted Duck | *Aythya fuligula* |
| Scaup | *Aythya marila* |
| Eider | *Somateria mollissima* |
| Long-tailed Duck | *Clangula hyemalis* |
| Common Scoter | *Melanitta nigra* |
| Velvet Scoter | *Mellanita fusca* |
| Goldeneye | *Bucephala clangula* |
| Smew | *Mergus albellus* |
| Red-breasted Merganser | *Mergus serrator* |
| Goosander | *Mergus merganser* |
| Ruddy Duck | *Oxyura jamaicensis* |
| Black Kite | *Milvus migrans* |
| Red Kite | *Milvus milvus* |
| Marsh Harrier | *Circus aeruginosus* |
| Hen Harrier | *Circus cyaneus* |
| Montagu's Harrier | *Circus pygargus* |
| Goshawk | *Accipiter gentilis* |
| Sparrowhawk | *Accipiter nisus* |
| Buzzard | *Buteo buteo* |
| Rough-legged Buzzard | *Buteo lagopus* |
| Osprey | *Pandion haliaetus* |
| Lesser Kestrel | *Falco naumanni* |
| Kestrel | *Falco tinnunculus* |
| Red-footed Falcon | *Falco vespertinus* |
| Merlin | *Falco columbarius* |
| Hobby | *Falco subbuteo* |
| Peregrine | *Falco peregrinus* |
| Red Grouse | *Lagopus lagopus* |
| Black Grouse | *Tetrao tetrix* |
| Red-legged Partridge | *Alectoris rufa* |
| Grey Partridge | *Perdix perdix* |
| Quail | *Coturnix coturnix* |
| Pheasant | *Phasianus colchicus* |
| Water Rail | *Rallus aquaticus* |
| Spotted Crake | *Porzana porzana* |
| Moorhen | *Gallinula chloropus* |
| Coot | *Fulica atra* |
| Oystercatcher | *Haematopus ostralegus* |
| Avocet | *Recurvirostra avosetta* |
| Little Ringed Plover | *Charadrius dubius* |
| Ringed Plover | *Charadrius hiaticula* |
| Killdeer | *Charadrius vociferus* |
| Kentish Plover | *Charadrius alexandrinus* |
| Dotterel | *Charadrius morinellus* |
| Golden Plover | *Pluvialis apricaria* |
| Grey Plover | *Pluvialis squatarola* |
| Lapwing | *Vanellus vanellus* |
| Knot | *Calidris canutus* |
| Sanderling | *Calidris alba* |
| Little Stint | *Calidris minuta* |
| Temminck's Stint | *Calidris temminckii* |
| Least Sandpiper | *Calidris minutilla* |
| White-rumped Sandpiper | *Calidris fuscicollis* |
| Pectoral Sandpiper | *Calidris melanotos* |
| Curlew Sandpiper | *Calidris ferruginea* |
| Purple Sandpiper | *Calidris maritima* |
| Dunlin | *Calidris alpina* |

| English Name | Scientific Name |
|---|---|
| Broad-billed Sandpiper | *Limicola falcinellus* |
| Buff-breasted Sandpiper | *Tryngites subruficollis* |
| Ruff | *Philomachus pugnax* |
| Jack Snipe | *Lymnocryptes minimus* |
| Snipe | *Gallinago gallinago* |
| Long-billed Dowitcher | *Limnodromus scolopaceus* |
| Woodcock | *Scolopax rusticola* |
| Black-tailed Godwit | *Limosa limosa* |
| Bar-tailed Godwit | *Limosa lapponica* |
| Whimbrel | *Numenius phaeopus* |
| Curlew | *Numenius arquata* |
| Spotted Redshank | *Tringa erythropus* |
| Redshank | *Tringa totanus* |
| Greenshank | *Tringa nebularia* |
| Lesser Yellowlegs | *Tringa flavipes* |
| Green Sandpiper | *Tringa ochropus* |
| Wood Sandpiper | *Tringa glareola* |
| Common Sandpiper | *Actitis hypoleucos* |
| Spotted Sandpiper | *Actitis macularia* |
| Turnstone | *Arenaria interpres* |
| Wilson's Phalarope | *Phalaropus tricolor* |
| Red-necked Phalarope | *Phalaropus lobatus* |
| Grey Phalarope | *Phalaropus fulicarius* |
| Pomarine Skua | *Stercorarius pomarinus* |
| Arctic Skua | *Stercorarius parasiticus* |
| Great Skua | *Stercorarius skua* |
| Mediterranean Gull | *Larus melanocephalus* |
| Laughing Gull | *Larus atricilla* |
| Little Gull | *Larus minutus* |
| Sabine's Gull | *Larus sabini* |
| Black-headed Gull | *Larus ridibundus* |
| Ring-billed Gull | *Larus delawarensis* |
| Common Gull | *Larus canus* |
| Lesser Black-backed Gull | *Larus fuscus* |
| Herring Gull | *Larus argentatus* |
| Iceland Gull | *Larus glaucoides* |
| Glaucous Gull | *Larus hyperboreus* |
| Great Black-backed Gull | *Larus marinus* |
| Kittiwake | *Rissa tridactyla* |
| Caspian Tern | *Sterna caspia* |
| Sandwich Tern | *Sterna sandvicensis* |
| Roseate Tern | *Sterna dougallii* |
| Common Tern | *Sterna hirundo* |
| Arctic Tern | *Sterna paradisaea* |
| Little Tern | *Sterna albifrons* |
| Black Tern | *Chlidonias niger* |
| White-winged Black Tern | *Chlidonias leucopterus* |
| Little Auk | *Alle alle* |
| Stock Dove | *Columba oenas* |
| Woodpigeon | *Columba palumbus* |
| Collared Dove | *Streptopelia decaocto* |
| Turtle Dove | *Streptopelia turtur* |
| Cuckoo | *Cuculus canorus* |
| Barn Owl | *Tyto alba* |
| Snowy Owl | *Nyctea scandiaca* |
| Little Owl | *Athene noctua* |
| Tawny Owl | *Strix aluco* |
| Long-eared Owl | *Asio otus* |

# LIST OF BIRDS MENTIONED IN TEXT

| English Name | Scientific Name |
|---|---|
| Short-eared Owl | *Asio flammeus* |
| Nightjar | *Caprimulgus europaeus* |
| Swift | *Apus apus* |
| Kingfisher | *Alcedo atthis* |
| Hoopoe | *Upupa epops* |
| Wryneck | *Jynx torquilla* |
| Green Woodpecker | *Picus viridis* |
| Great Spotted Woodpecker | *Dendrocopos major* |
| Lesser Spotted Woodpecker | *Dendrocopos minor* |
| Woodlark | *Lullula arborea* |
| Skylark | *Alauda arvensis* |
| Sand Martin | *Riparia riparia* |
| Swallow | *Hirundo rustica* |
| Red-rumped Swallow | *Hirundo daurica* |
| House Martin | *Delichon urbica* |
| Tree Pipit | *Anthus trivialis* |
| Meadow Pipit | *Anthus pratensis* |
| Rock Pipit | *Anthus petrosus* |
| Water Pipit | *Anthus spinoletta* |
| Yellow Wagtail | *Motacilla flava flavissima* |
|    Blue-headed Wagtail | *Motacilla flava flava* |
| Grey Wagtail | *Motacilla cinerea* |
| Pied Wagtail | *Motacilla alba yarrellii* |
|    White Wagtail | *Motacilla alba alba* |
| Dipper | *Cinclus cinclus* |
| Wren | *Troglodytes troglodytes* |
| Dunnock | *Prunella modularis* |
| Robin | *Erithacus rubecula* |
| Nightingale | *Luscinia megarhynchos* |
| Bluethroat | *Luscinia svecica* |
| Black Redstart | *Phoenicurus ochruros* |
| Redstart | *Phoenicurus phoenicurus* |
| Whinchat | *Saxicola rubetra* |
| Stonechat | *Saxicola torquata* |
| Wheatear | *Oenanthe oenanthe* |
| Ring Ouzel | *Turdus torquatus* |
| Blackbird | *Turdus merula* |
| Fieldfare | *Turdus pilaris* |
| Song Thrush | *Turdus philomelos* |
| Redwing | *Turdus iliacus* |
| Mistle Thrush | *Turdus viscivorus* |
| Cetti's Warbler | *Cettia cetti* |
| Grasshopper Warbler | *Locustella naevia* |
| Savi's Warbler | *Locustella luscinioides* |
| Sedge Warbler | *Acrocephalus schoeobaenus* |
| Marsh Warbler | *Acrocephalus palustris* |
| Reed Warbler | *Acrocephalus scirpaceus* |
| Great Reed Warbler | *Acrocephalus arundinaceus* |
| Barred Warbler | *Sylvia nisoria* |
| Lesser Whitethroat | *Sylvia curruca* |
| Whitethroat | *Sylvia communis* |
| Garden Warbler | *Sylvia borin* |
| Blackcap | *Sylvia atricapilla* |
| Yellow-browed Warbler | *Phylloscopus inornatus* |
| Wood Warbler | *Phylloscopus sibilatrix* |
| Chiffchaff | *Phylloscopus collybita* |
| Willow Warbler | *Phylloscopus trochilus* |
| Goldcrest | *Regulus regulus* |

| English Name | Scientific Name |
|---|---|
| Firecrest | *Regulus ignicapillus* |
| Spotted Flycatcher | *Muscicapa striata* |
| Pied Flycatcher | *Ficedula hypoleuca* |
| Bearded Tit | *Panurus biarmicus* |
| Long-tailed Tit | *Aegithalos caudatus* |
| Marsh Tit | *Parus palustris* |
| Willow Tit | *Parus montanus* |
| Coal Tit | *Parus ater* |
| Blue Tit | *Parus caeruleus* |
| Great Tit | *Parus major* |
| Nuthatch | *Sitta europaea* |
| Treecreeper | *Certhia familiaris* |
| Golden Oriole | *Oriolus oriolus* |
| Red-backed Shrike | *Lanius collurio* |
| Great Grey Shrike | *Lanius excubitor* |
| Jay | *Garrulus glandarius* |
| Magpie | *Pica pica* |
| Chough | *Pyrrhocorax pyrrhocorax* |
| Jackdaw | *Corvus monedula* |
| Rook | *Corvus frugilegus* |
| Carrion Crow | *Corvus corone* |
| Raven | *Corvus corax* |
| Starling | *Sturnus vulgaris* |
| House Sparrow | *Passer domesticus* |
| Tree Sparrow | *Passer montanus* |
| Chaffinch | *Fringilla coelebs* |
| Brambling | *Fringilla montifringilla* |
| Greenfinch | *Carduelis chloris* |
| Goldfinch | *Carduelis carduelis* |
| Siskin | *Carduelis spinus* |
| Linnet | *Carduelis cannabina* |
| Twite | *Carduelis flavirostris* |
| Redpoll | *Carduelis flammea* |
| Two-barred Crossbill | *Loxia leucoptera* |
| Crossbill | *Loxia curvirostra* |
| Bullfinch | *Pyrrhula pyrrhula* |
| Hawfinch | *Coccothraustes coccothraustes* |
| Lapland Bunting | *Calcarius lapponicus* |
| Snow Bunting | *Plectrophenax nivalis* |
| Yellowhammer | *Emberiza spodocephala* |
| Reed Bunting | *Emberiza schoeniclus* |
| Corn Bunting | *Miliaria calandra* |

# COMPARISON OF SITES

To help readers to decide where to visit, we have analysed sites by their main habitats. The order of habitats is based on that adopted by the Nature Conservancy Council (NCC) in *A Nature Conservation Review*. Within each habitat sites are listed alphabetically and rated according to our assessment of their quality. This assessment descends from three stars (***), which denotes consistently good sites, to one star (*), for areas of more limited, local interest. These assessments are relative only to other sites in the West Midlands. No attempt has been made at a national assessment.

## Deciduous/Mixed Woodland

- ** Bannister's Coppice
- * Belvide
- ** Bentley and Monks Park
- * Bittell
- * Blithfield
- * Bredon Hill
- * Bromwich Wood
- * Brown Clee Hill
- ** Cannock Chase
- * Chelmarsh Reservoir
- ** Chillington Avenue
- *** Churnet
- * Colstey Wood
- ** Dales
- * Doward, The
- *** Earl's Hill
- ** Ercall Woods
- * Eywood
- * Hartshill Hayes
- ** Haughmond Hill
- * Himley and Baggeridge
- * Knapp, The
- ** Lickey Hills
- * Malvern Hills
- * Mary Knoll Valley
- ** North Shropshire Meres (especially Blake Mere, Cole Mere and Ellesmere)
- ** North Staffordshire Moors
- ** Olchon Valley
- * Pipers Hill
- ** Rectory Wood (Long Mynd)
- * Rudyard
- * Ryton Wood
- * Saltwells Wood
- * Sandwell Valley
- *** Severn Gorge (including Benthall Edge Wood)
- * Stiperstones
- * Stretton Hills
- * Sutton Park
- * Telford District
- ** Trench Wood
- * Trentham
- * Ufton Fields
- ** Wappenbury Wood
- ** Wenlock Edge
- ** Whitcliffe Common
- ** Woolhope Area
- * Wrekin Woods
- *** Wyre Forest

## Coniferous Woodland

- * Bentley and Monks Park
- * Brown Clee Hill
- *** Cannock Chase
- * Catherton Common
- ** Churnet Valley
- *** Colstey Wood
- * Hartshill Hayes
- * Haughmond Hill
- ** Haugh Wood
- * Lickey Hills
- * Long Mynd
- *** Mortimer Forest (Ludlow area and Wigmore Rolls)
- * Park Hall
- * Pontesford Hill
- * Stiperstones
- * Sutton Park
- * Wenlock Edge
- * Wrekin
- *** Wyre Forest

## Grassland and Scrub

- ** Brandon Marsh
- ** Bredon Hill
- * Brierley Hill Pools
- ** Brown Clee Hill
- ** Chasewater
- * Chelmarsh
- ** Earl's Hill
- ** Ladywalk
- * Lickey Hills
- ** Long Mynd
- ** Kingsbury Water Park
- *** Malvern Hills
- * Mary Knoll Valley

213

* Park Hall
* Saltwells Wood
** Sandwell Valley
* Stiperstones
** Stretton Hills
** Sutton Park
* Telford area
* Titterstone Clee Hill
* Ufton Fields
* Wrekin

## Heathland

* Brown Clee Hill
*** Cannock Chase
* Castlemorton Common
** Catherton Common
* Chasewater
* Haughmond Hill
** Sutton Park
* Titterstone Clee Hill

## Lakes, Pools and Reservoirs

** Alvecote Pools
* Bartley
*** Belvide
** Bittell
*** Blithfield
** Brandon Marsh
* Brierley Hill Pools
*** Chasewater
** Chelmarsh Reservoir
** Coton
*** Draycote
* Edgbaston Reservoir
* Eywood Pool
* Himley
*** Kingsbury Water Park
** Lea Marston
* Longsdon Mill Pool
* Middleton Pool
*** North Shropshire Meres
* Rudyard
* Ryton Pool
** Sandwell Valley
* Shustoke
* Sutton Park
* Telford Pools
** Tittesworth
* Trentham Park
* Ufton Fields
*** Upton Warren
** Westport Lake

## Rivers and Streams

* Alvecote Pools
** Avon Valley
* Blackbrook Valley
* Churnet
** Coombes Valley
** Dovedale
*** Dowles Brook
** Earl's Hill
** Hamps
** Hereford Wye and Lugg
** Knapp, The
** Long Mynd
** Manifold
** Middle Wye
* Olchon Brook
** River Severn
** River Teme
* Sheinton Brook
* Tame Valley
* Wye Gorge

## Flood Meadows

** Avon Valley
** Doxey Marshes
* Hereford Wye and Lugg
** Middle Wye
** Severn Valley
* Tame Valley
** Teme Valley

## Marsh

* Belvide
*** Brandon Marsh
* Chelmarsh Reservoir
*** Doxey Marshes
* Ford Green
** Kingsbury Water Park
** Ladywalk
* Nether Whitacre
* Upton Warren

## Moorland

** Black Mountains
** Brown Clee Hill
*** Long Mynd
*** North Staffordshire Moors
** Stiperstones
* Titterstone Clee Hill

# MAPS

# Key to Sites

## Map 1: Principal Birdwatching Sites

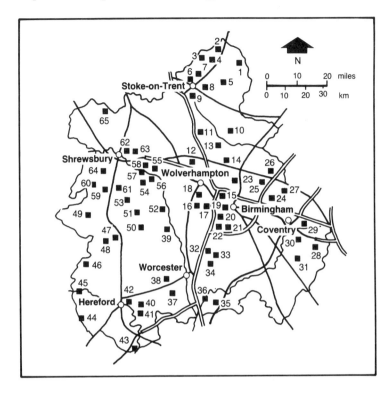

## Key to Maps

| | | | |
|---|---|---|---|
| Motorways and junctions | | Built-up areas | |
| Class A and B roads | | Deciduous and mixed woodland | |
| Other roads | | Coniferous woodland | |
| Unmetalled tracks | | Water | |
| Car parks | | Marsh | |
| Footpaths | | Sand or mud | |
| Railways and stations | | Steep slopes | |
| Canals | | Site boundaries (where appropriate) | |
| Main rivers | | Hides | |
| Other rivers and streams | | Triangulation point | |

## Map 2 : The Dales

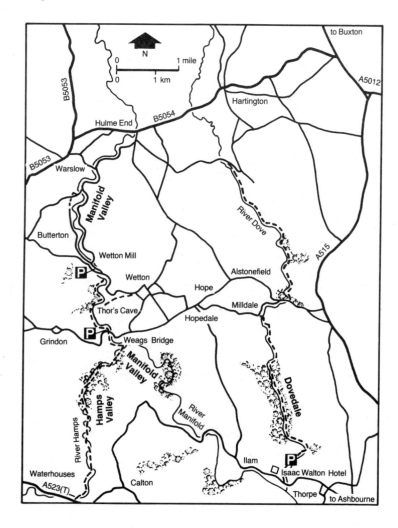

## Map 3 : The North Staffordshire Moors

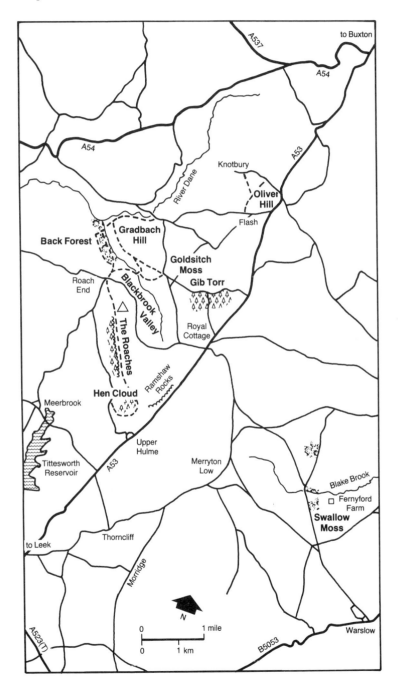

# Map 4 : Longsdon, Rudyard and Tittesworth

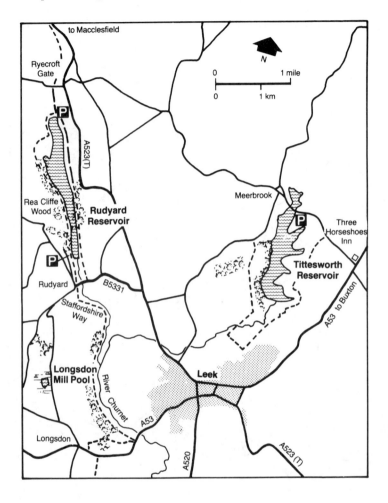

## Map 5 : The Churnet Valley

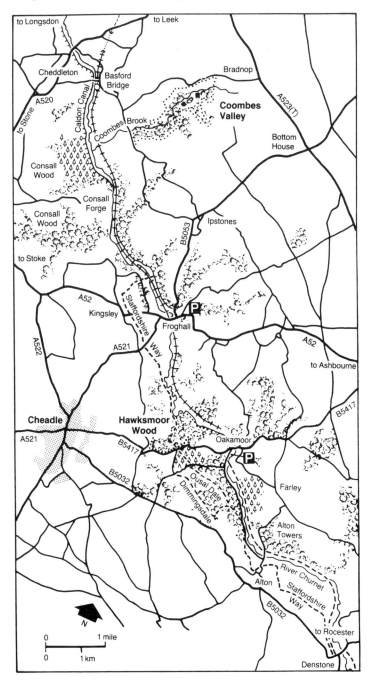

# Map 6 : Westport Lake and Ford Green

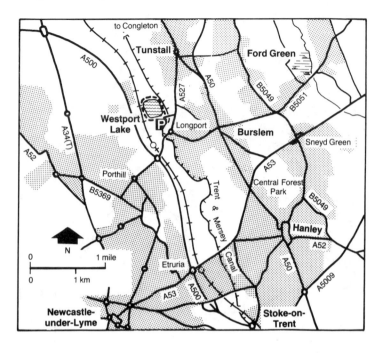

## Map 7 : Park Hall and Trentham Park

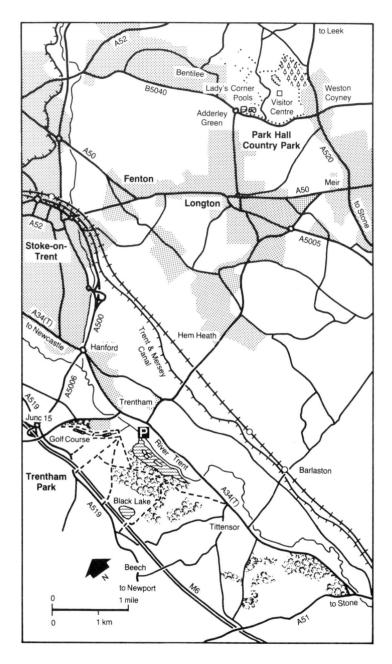

# Map 8 : Blithfield Reservoir

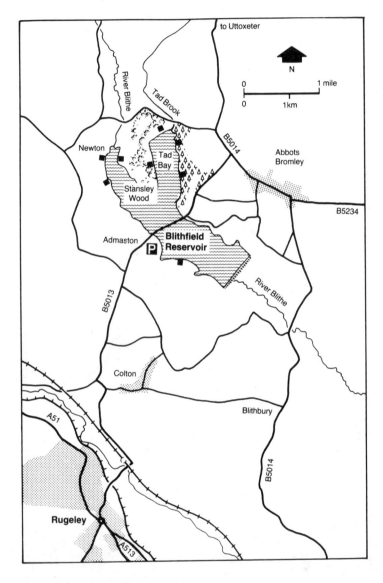

# Map 9 : Doxey Marshes

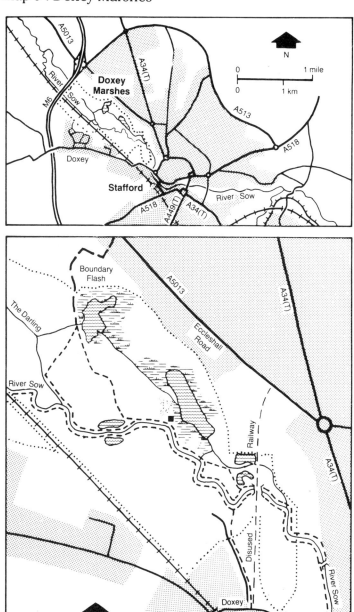

# Map 10 : Belvide Reservoir and Chillington Lower Avenue

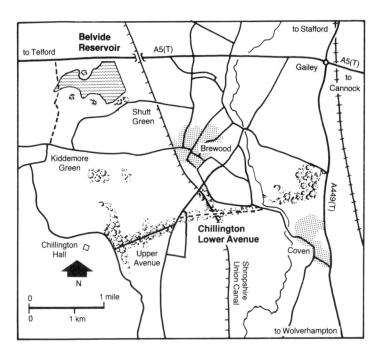

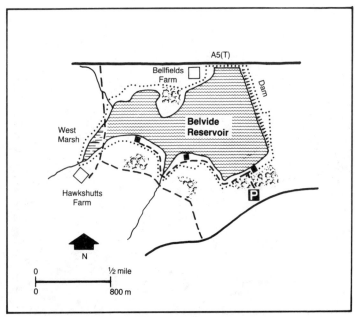

# Map 11 : Cannock Chase

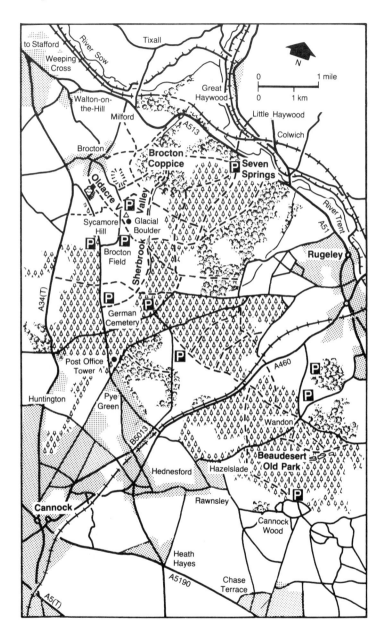

# Map 12 : Chasewater

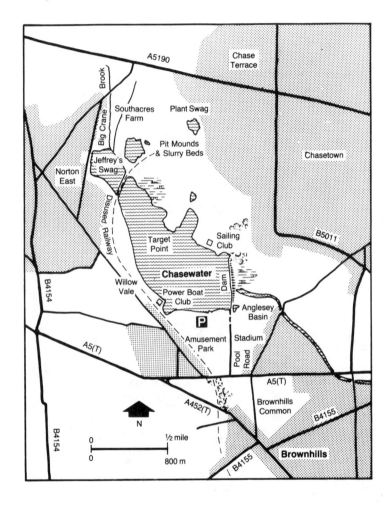

## Map 13 : Sandwell Valley

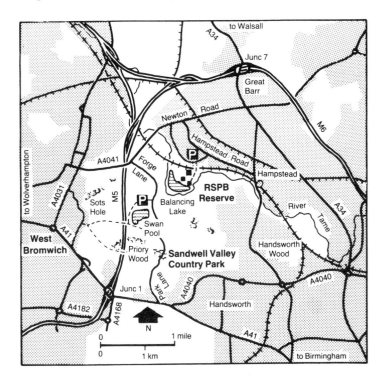

## Map 14 : Brierley Hill Pools, Saltwells Wood and Himley

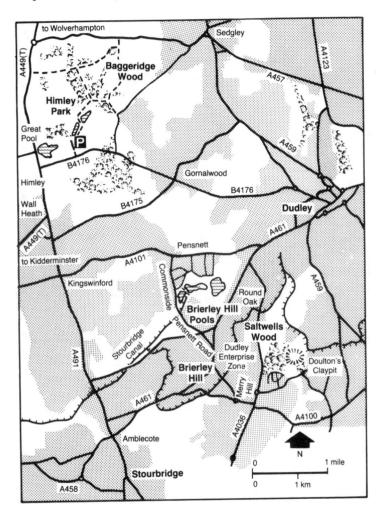

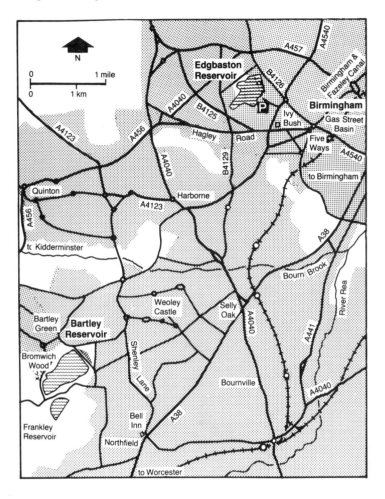

## Map 15 : Edgbaston and Bartley Reservoirs

## Map 16 : Bittell Reservoir and Lickey Hills

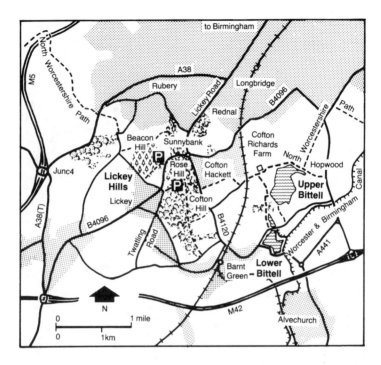

## Map 17: Sutton Park

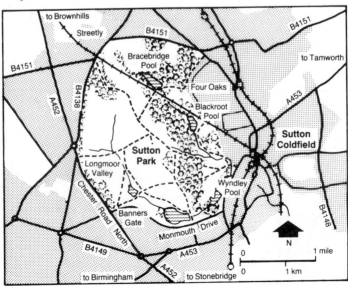

# Map 18 : The Tame Valley

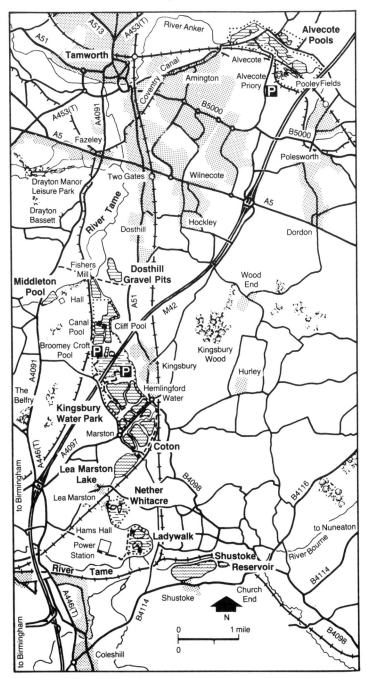

# Map 19 : Bentley Park, Monks Park and Hartshill Hayes

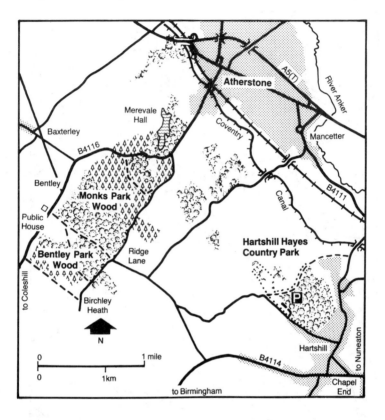

## Map 20 : Draycote Water

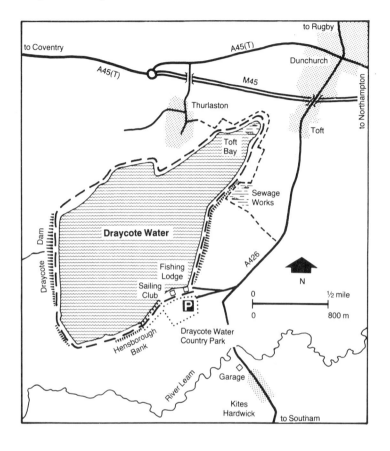

# Map 21 : Brandon Marsh

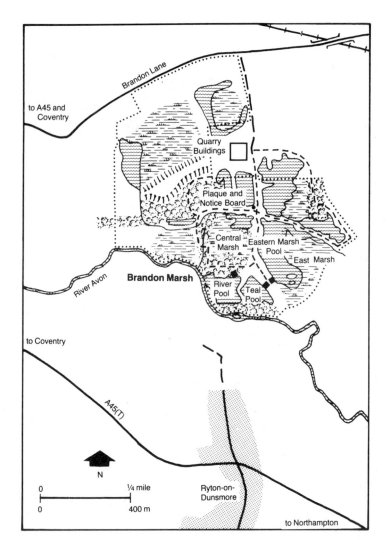

# Map 22 : Ryton and Wappenbury

# Map 23 : Ufton Fields

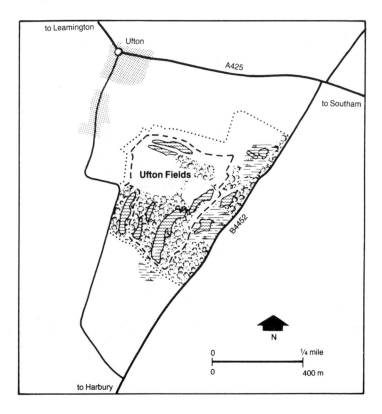

# Map 24 : Upton Warren

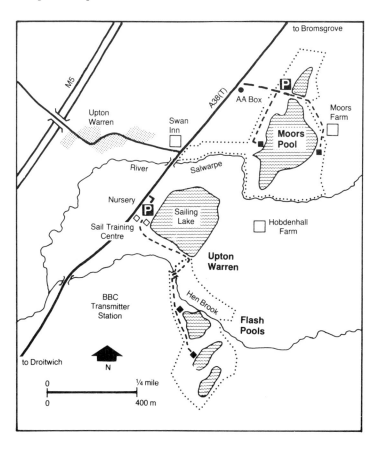

## Map 25 : Pipers Hill and Trench Wood

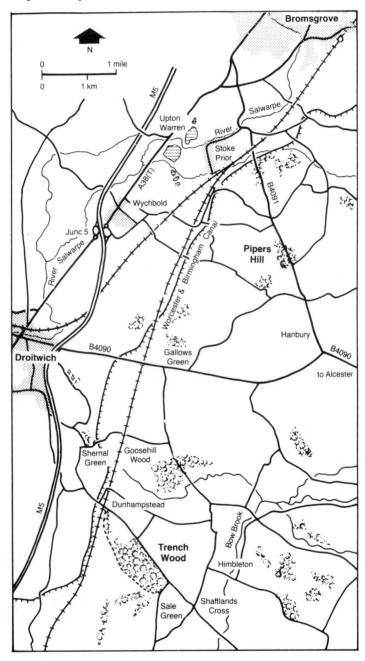

## Map 26 : Bredon Hill and The Avon Valley

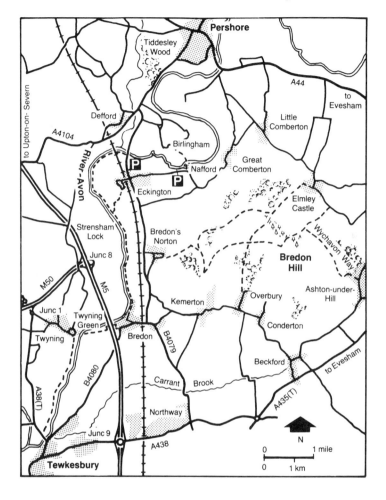

## Map 27 : The Malvern Hills

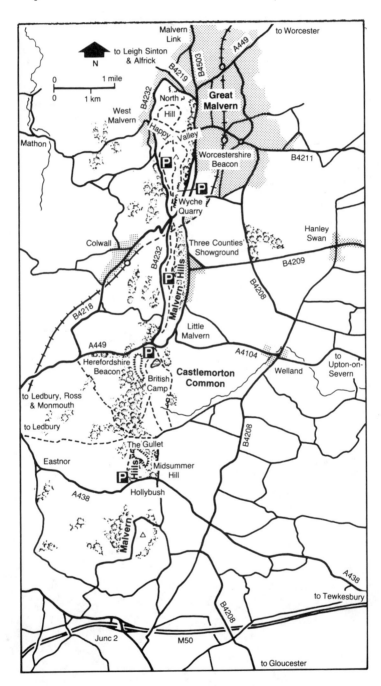

## Map 27a : The Knapp and Papermill

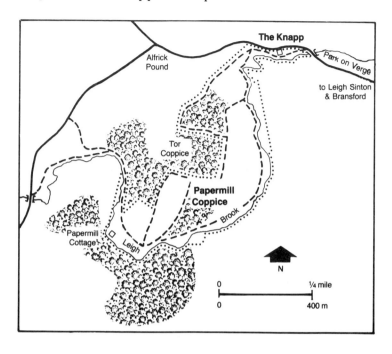

**The Knapp**

Park on Verge

Alfrick
Pound

to Leigh Sinton
& Bransford

Tor
Coppice

**Papermill
Coppice**

Brook

Papermill
Cottage

Leigh

N

0             ¼ mile

0             400 m

## Map 28 : Wyre Forest

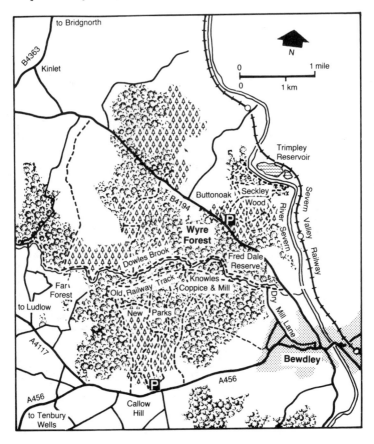

## Map 29 : The Woolhope Area

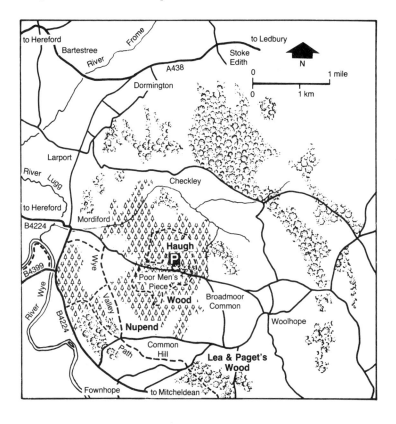

## Map 30 : Hereford Lugg and Wye

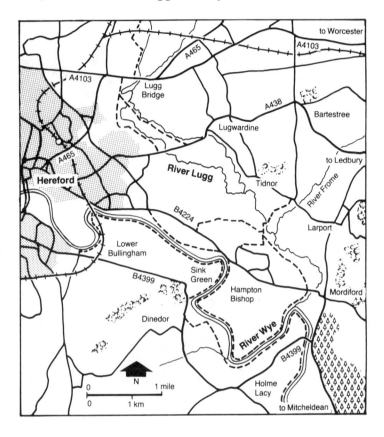

## Map 31 : Wye Gorge

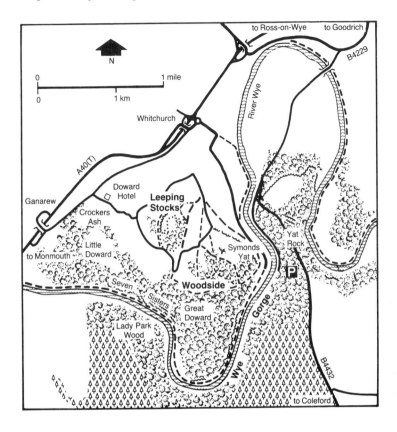

## Map 32 : The Black Mountains

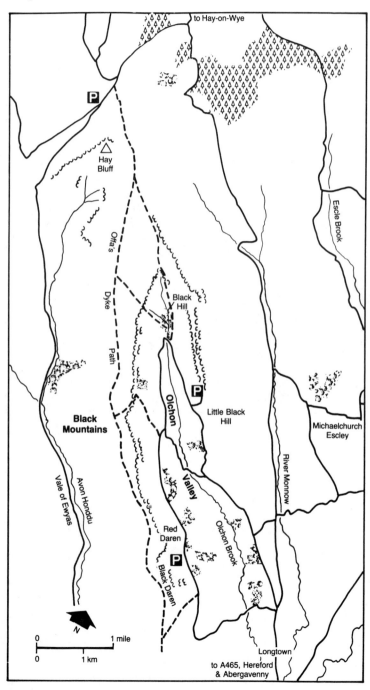

Map 33 : The Middle Wye Valley

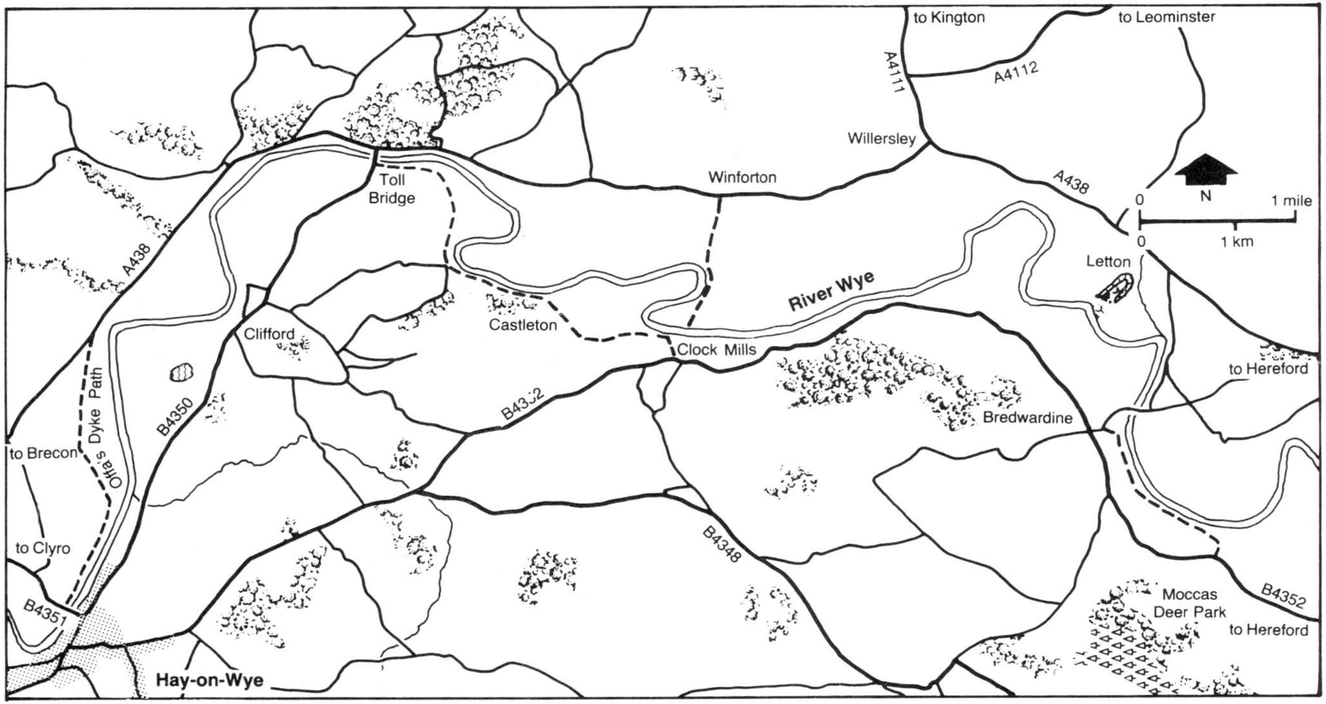

# Map 34 : Eywood Pool

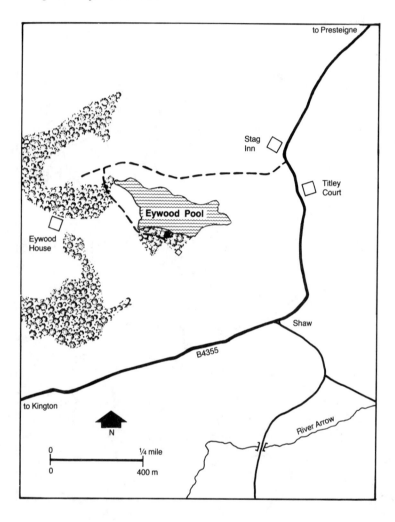

## Map 35 : Whitcliffe Common and Mary Knoll Valley

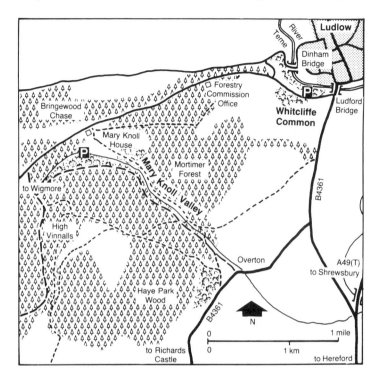

## Map 36 : Wigmore Area

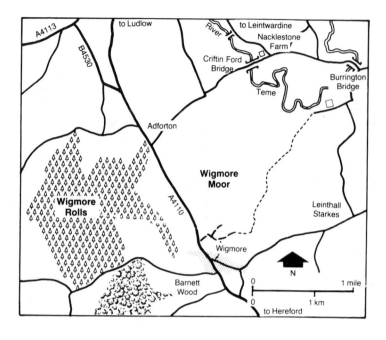

## Map 37 : Colstey Wood and Bury Ditches

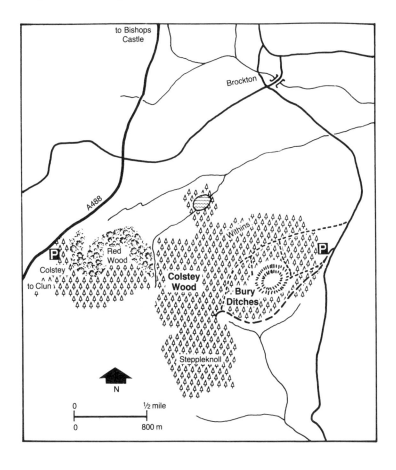

## Map 38 : Brown Clee Hill

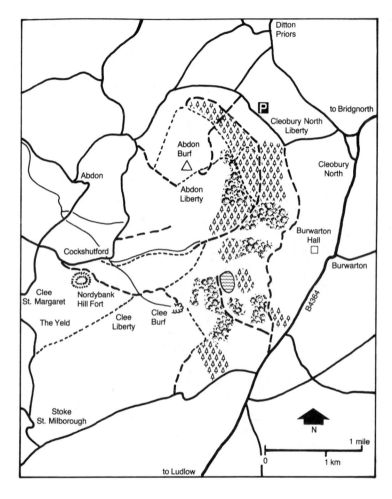

## Map 39 : Titterstone Clee Hill and Catherton Common

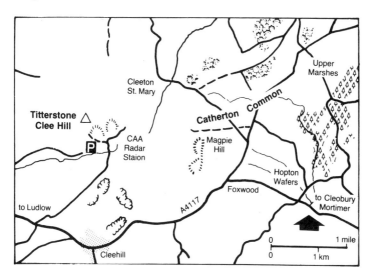

## Map 40 : Chelmarsh Reservoir and the River Severn

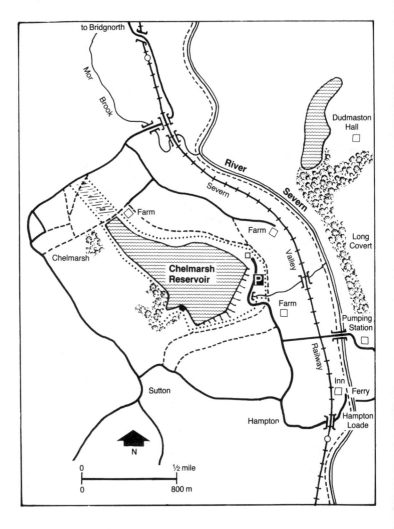

## Map 41 : Wenlock Edge

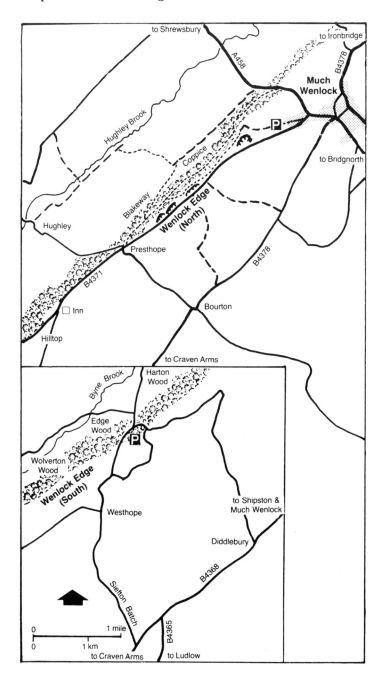

# Map 41a : Bannister's Coppice

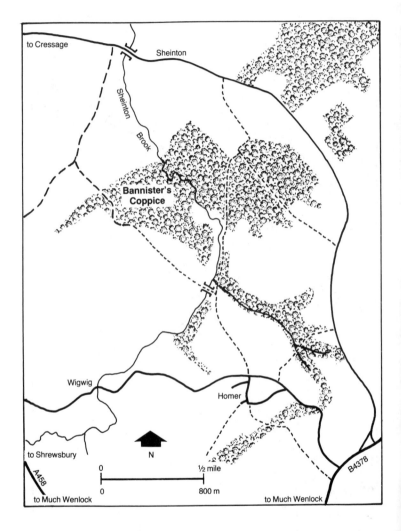

# Map 42: The Telford Area

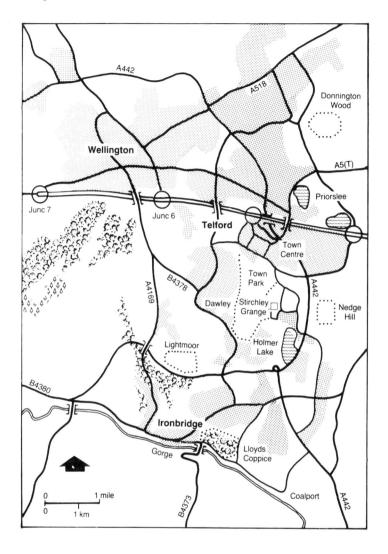

## Map 43 : Benthall Edge Wood

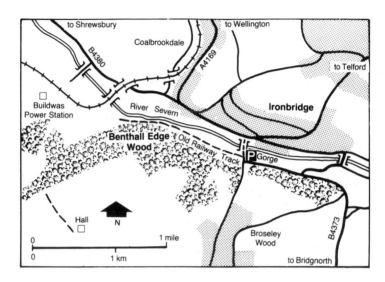

## Map 44 : River Severn : Cressage to Buildwas

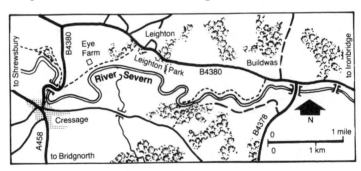

## Map 45 : The Wrekin and The Ercall

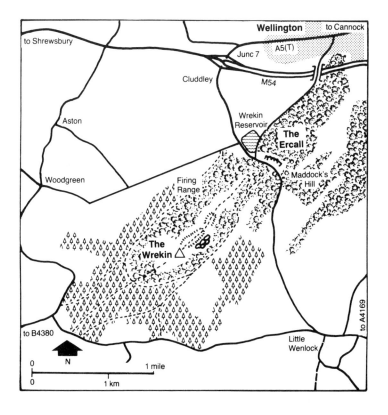

# Map 46 : Long Mynd

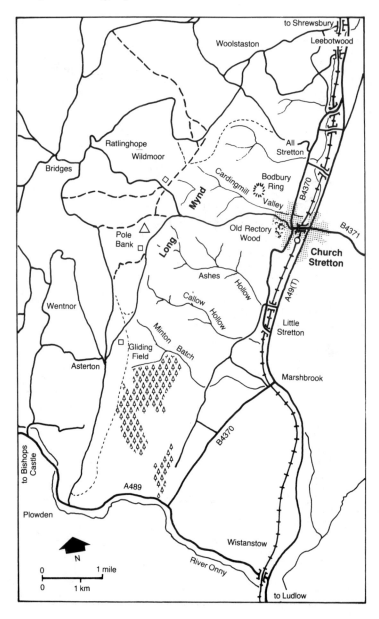

# Map 47 : Stiperstones

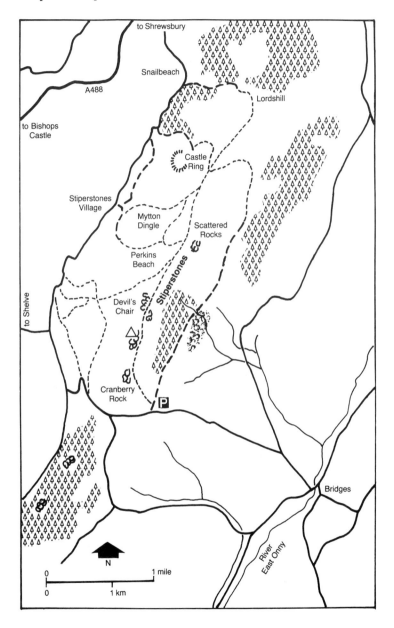

to Shrewsbury

Snailbeach

A488

Lordshill

to Bishops
Castle

Castle
Ring

Stiperstones
Village

Mytton
Dingle

Scattered
Rocks

Perkins
Beach

Stiperstones

to Shelve

Devil's
Chair

Cranberry
Rock

P

Bridges

River
East Onny

N

0 _____ 1 mile

0 _____ 1 km

## Map 48 : The Stretton Hills

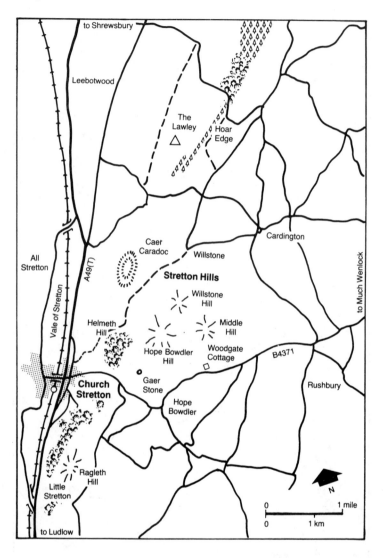

## Map 49 : Monkmoor and Haughmond Hill

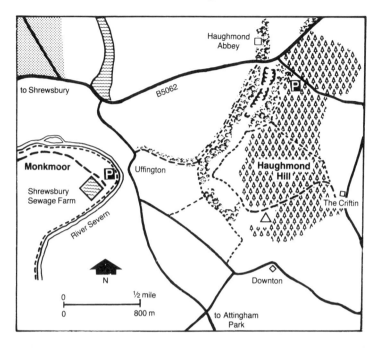

to Shrewsbury

B5062

Haughmond
Abbey

P

Monkmoor

P

Uffington

Haughmond
Hill

Shrewsbury
Sewage Farm

The Criftin

River Severn

N

Downton

0           ½ mile

0           800 m

to Attingham
Park

# Map 50 : Earl's Hill and Pontesford Hill

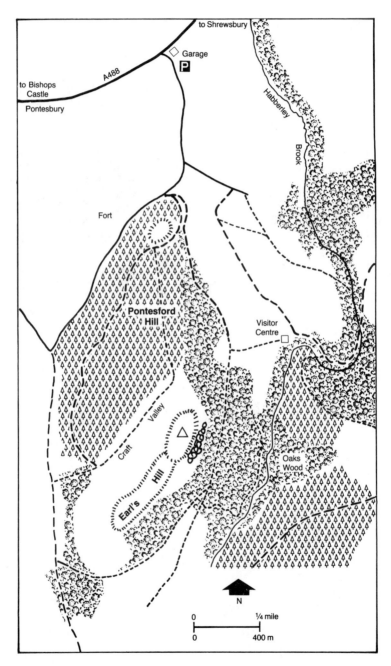

## Map 51 : The North Shropshire Meres

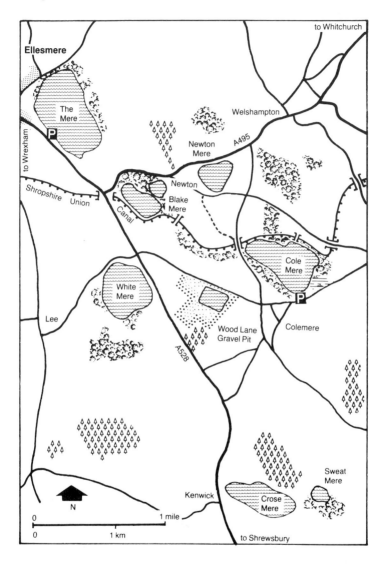

# INDEX